Springer Theses

Recognizing Outstanding Ph.D. Research

For further volumes:
http://www.springer.com/series/8790

Aims and Scope

The series "Springer Theses" brings together a selection of the very best Ph.D. theses from around the world and across the physical sciences. Nominated and endorsed by two recognized specialists, each published volume has been selected for its scientific excellence and the high impact of its contents for the pertinent field of research. For greater accessibility to non-specialists, the published versions include an extended introduction, as well as a foreword by the student's supervisor explaining the special relevance of the work for the field. As a whole, the series will provide a valuable resource both for newcomers to the research fields described, and for other scientists seeking detailed background information on special questions. Finally, it provides an accredited documentation of the valuable contributions made by today's younger generation of scientists.

Theses are accepted into the series by invited nomination only and must fulfill all of the following criteria

- They must be written in good English.
- The topic should fall within the confines of Chemistry, Physics, Earth Sciences, Engineering and related interdisciplinary fields such as Materials, Nanoscience, Chemical Engineering, Complex Systems and Biophysics.
- The work reported in the thesis must represent a significant scientific advance.
- If the thesis includes previously published material, permission to reproduce this must be gained from the respective copyright holder.
- They must have been examined and passed during the 12 months prior to nomination.
- Each thesis should include a foreword by the supervisor outlining the significance of its content.
- The theses should have a clearly defined structure including an introduction accessible to scientists not expert in that particular field.

Ryuji Okazaki

Hidden Order and Exotic Superconductivity in the Heavy-Fermion Compound URu_2Si_2

Doctoral Thesis accepted by
Kyoto University, Kyoto, Japan

Author
Dr. Ryuji Okazaki
Department of Physics
Nagoya University
Nagoya
Japan

Supervisor
Prof. Yuji Matsuda
Department of Physics
Kyoto University
Kyoto
Japan

ISSN 2190-5053 ISSN 2190-5061 (electronic)
ISBN 978-4-431-56359-4 ISBN 978-4-431-54592-7 (eBook)
DOI 10.1007/978-4-431-54592-7
Springer Tokyo Heidelberg New York Dordrecht London

Softcover reprint of the hardcover 1st edition 2013

Printed on acid-free paper

Springer is part of Springer Science+Business Media (www.springer.com)

Parts of this thesis have been published in the following journals:

1. **R. Okazaki**, T. Shibauchi, H. J. Shi, Y. Haga, T. D. Matsuda, E. Yamamoto, Y. Onuki, H. Ikeda, and Y. Matsuda; "Rotational Symmetry Breaking in the Hidden-Order Phase of URu_2Si_2" Science **331**, 439 (2011).
2. **R. Okazaki**, M. Shimozawa, H. Shishido, M. Konczykowski, Y. Haga, T. D. Matsuda, E. Yamamoto, Y. Onuki, Y. Yanase, T. Shibauchi, and Y. Matsuda; "Anomalous Temperature Dependence of Lower Critical Field in Ultraclean URu_2Si_2" J. Phys. Soc. Jpn. **79**, 084705 (2010).
3. **R. Okazaki**, M. Konczykowski, C. J. van der Beek, T. Kato, K. Hashimoto, M. Shimozawa, H. Shishido, M. Yamashita, M. Ishikado, H. Kito, A. Iyo, H. Eisaki, S. Shamoto, T. Shibauchi, and Y. Matsuda; "Lower critical fields of superconducting $PrFeAsO_{1-y}$ single crystals" Phys. Rev. B **79**, 064520 (2009).
4. **R. Okazaki**, Y. Kasahara, H. Shishido, M. Konczykowski, K. Behnia, Y. Haga, T. D. Matsuda, Y. Onuki, T. Shibauchi, and Y. Matsuda; "Flux Line Lattice Melting and the Formation of a Coherent Quasiparticle Bloch State in the Ultraclean URu_2Si_2 Superconductor" Phys. Rev. Lett. **100**, 037004 (2008).
5. **R. Okazaki**, H. Shishido, T. Shibauchi, M. Konczykowski, A. Buzdin, and Y. Matsuda; "High-field superconducting transition of $CeCoIn_5$ studied by local magnetic induction measurements" Phys. Rev. B **76**, 224529 (2007).

Supervisor's Foreword

This Ph.D. thesis describes an investigation conducted by Dr. Ryuji Okazaki during his doctoral program at the Graduate School of Science, Kyoto University. During nearly 5 years at the graduate school, he achieved considerable results in research of the phase transition and superconductivity in heavy fermion materials, which belong to the group of strongly correlated electron systems. In particular, he made several important findings in the uranium-based heavy fermion compound URu_2Si_2.

As the Supervisor of his master's and doctoral courses, I can introduce the two most important findings in his studies. The first concerns a mysterious phase transition of URu_2Si_2, which is called hidden order transition. The development of collective long-range order via phase transitions occurs by the spontaneous breaking of fundamental symmetries. For example, magnetism is a consequence of broken time-reversal symmetry. The broken symmetry that develops below $T_0 = 17.5$ K in URu_2Si_2 has eluded such identification for more than a quarter of a century. By developing extremely sensitive magnetic measurements, he has unveiled what symmetry is broken in the hidden order phase. This result demonstrates that the hidden order phase is an electronic nematic phase, a translationally invariant metallic phase with spontaneous breaking of rotational symmetry. His findings provide a key to resolve the order parameter of the hidden order phase, which has been a longstanding mystery.

The second concerns another aspect of URu_2Si_2, i.e., the anomalous superconducting properties. The exotic superconducting state with unconventional pairing has been a central issue in condensed matter physics. Through high-precision measurements of lower critical field H_{c1} using ultraclean single crystals of URu_2Si_2, Dr. Okazaki found that the temperature dependence of H_{c1} significantly differs from that of any superconductors. He proposed that this behavior can be accounted for in terms of the multicomponent superconducting order parameter with broken time reversal symmetry. In addition, apart from the superconducting symmetry, he found that in ultraclean URu_2Si_2 a distinct flux lattice melting with outstanding characters occurs in the vortex state, which is the first demonstration of the melting at sub-Kelvin temperatures. These results demonstrate that URu_2Si_2 also provides a new "playground" for studying the novel superconducting state.

I hope that there will be many grateful readers who will gain a broader perspective of the phase transition and superconducting properties in these fascinating systems as a result of Dr. Okazaki's efforts.

Kyoto, Japan, October 2013 Prof. Yuji Matsuda

Acknowledgments

I would like to express my great appreciation to Professor Yuji Matsuda for his valuable advice, general support, and great encouragement throughout my Ph.D. studies. His immense perspective for research has exerted a strong influence on me. This study has been achieved with a great deal of his continued warm guidance.

I am sincerely grateful to Professor Takasada Shibauchi, who provided me with extensive advice as well as fruitful discussion on my experimental results. This work was improved with his helpful advice. I have also been influenced very much by his serious scientific attitude and way of thinking.

I appreciate the great encouragement and helpful discussion of Professor Ichiro Terasaki. His impressive mind for science has exerted a significant influence on me.

I also wish to express my appreciation to Professor Yuichi Kasahara for experimental advice and for providing me with valuable data on URu_2Si_2. I appreciate, as well, Professor Yoichi Yanase and Professor Hiroaki Ikeda for theoretical advice with regard to the hidden-order and superconducting states of URu_2Si_2. I would like to thank Professor Minoru Yamashita, Professor Yasuyuki Nakajima, Professor Hiroaki Shishido, and Professor Shigeru Kasahara for their experimental support, fruitful discussion, and strong encouragement.

I am also appreciative of the help received from Professor Yoshichika Onuki, Dr. Yoshinori Haga, Dr. Tatsuma Matsuda, and Dr. Etsuji Yamamoto for providing me with very-high-quality URu_2Si_2 single crystals. I would like to acknowledge Dr. Marcin Konczykowski and Dr. Kees van der Beek for their kind support during my stay at the École Polytechnique, for their technical advice, and for providing with many Hall probes.

I thank Professor Kenichiro Hashimoto for discussion on the penetration depth study, and I appreciate the assistance of Mr. Masaaki Shimozawa in lower critical field measurement. I also thank Mr. Hong Jie Shi for his helpful assistance in torque measurement. I am deeply indebted to all members of Professor Matsuda and Professor Shibauchi's group at Kyoto University, namely Dr. Manabu Izaki, Mr. Tomonari Kato, Mr. Takuya Iwasawa, Mr. Sho Tonegawa, and Mr. Yuta Mizukami, for their helpful discussion and great support.

I am also grateful to the staff members of the Research Center for Low Temperature and Material Science at Kyoto University for providing me with liquid helium and liquid nitrogen. I wish to acknowledge with gratitude the financial support provided by the Research Fellowship of the Japan Society for Promotion of Science for Young Scientists.

Finally I deeply appreciate my parents, my wife, my brother, and the rest of my family, who always supported and encouraged me in many ways.

Contents

Chapter 1
Introduction

Phase transition phenomenon into various intriguing ordered states is one of the most fundamental subjects penetrating into diverse fields of physics. Here, we introduce the symmetry breaking and the order parameter associated with the phase transition phenomena. In general, a high-temperature disordered phase possesses a high symmetry (for instance, symmetry group G_1) and then has a large entropy to minimize the free energy. With lowering temperature, the system undergoes a low-temperature ordered phase with the reduced symmetry (symmetry group $G_2 \subset G_1$) through a phase transition at the critical temperature T_c. The symmetry is reduced to its subgroups when the system passes through the transition temperatures with decreasing temperature.

Now a long-range order in a low-temperature phase is characterized by a finite value of the correlation function between a quantity $A^\dagger$ measured at the origin ($r = 0$) and A at infinite distance ($r \to \infty$), which is expressed as

$$\lim_{r\to\infty} \langle A^\dagger(0)A(r)\rangle \neq 0. \tag{1.1}$$

This equation is satisfied by nonzero value of $\langle A(r)\rangle$. Here, $\langle A(r)\rangle$ is introduced as an order parameter, which shows a finite value in the low-temperature phase with the long-range order. In the high-temperature high-symmetry phase without the long-range order, on the other hand, the correlation function and the order parameter become zero. Thus, the order parameter can be regarded as a variable that describes the symmetry breaking in the system.

Let us look at a ferromagnetic order in solids. The Hamiltonian H for the three-dimensional ferromagnetic Heisenberg system is expressed as

$$H = J \sum_{\langle i,j\rangle} \mathbf{S}_i \cdot \mathbf{S}_j, \tag{1.2}$$

R. Okazaki, *Hidden Order and Exotic Superconductivity in the Heavy-Fermion Compound URu_2Si_2*, Springer Theses, DOI: 10.1007/978-4-431-54592-7_1,

where $J(< 0)$ is the exchange energy and $\langle i, j \rangle$ denotes the nearest–neighbor lattice-site pair. This Hamiltonian is invariant with respect to the rotational transformation for spin operator $\mathbf{S}$, thus it possesses the three-dimensional rotational symmetry $SO(3)$ as well as a translational symmetry due to the lattice periodicity. At high temperatures, the spin orientations are random, then the system has the high symmetry $G_1 = SO(3)$ (Fig. 1.1a). Below the transition temperature T_c, on the other hand, the spins point in a certain direction (Fig. 1.1b), then the system has the reduced symmetry $G_2 = U(1)(\subset G_1)$, which expresses the two-dimensional rotational symmetry around the spin-aligned direction. Here, such spin alignment produces a finite value of $\langle \mathbf{S}(\mathbf{r}) \rangle$, which is zero in the high-temperature disordered phase. Thus, the magnetization $\mathbf{M} = \langle \mathbf{S}(\mathbf{r}) \rangle$ is the order parameter for the ferromagnetic order.

Now the ferromagnetic order is described by the quenched spin degree of freedom of electrons at low temperatures. In condensed matter physics, electrons possess the multiple degrees of freedom such as charge, spin and orbital degrees of freedom, each of which exhibits various ordered states at low temperatures. Moreover, electron–electron interaction as well as the electron–phonon interaction often couple with such internal degrees of freedom, leading to rich electronic and magnetic phases in strongly correlated electron systems. However, the nature of several exotic ordered states, such as orbital order [1] or higher-rank multipole order [2, 3], is difficult to be elucidated. The most fundamental issue to address such problems is to identify which symmetry is broken and what is order parameter in such ordered phase.

In this study, we investigate the nature of mysterious phase transition phenomena in the heavy-fermion compound URu_2Si_2. This material exhibits two clear phase transitions at $T_0 = 17.5$ K and $T_c = 1.4$ K [4–6], as shown in Fig. 1.2. Both transitions are of second order, characterized by the distinct jump of specific heat. The former transition is called "hidden order", because what is the order parameter is still an open question since its discovery in 1985. The latter is a superconducting phase transition, but the superconducting symmetry, which is closely related to the pairing mechanism, is not identified. In spite of much experimental and theoretical effort, however, the nature of both exotic phases remains to be clarified. The purpose of our study is to investigate the nature of both hidden-order and superconducting phases, especially the detailed knowledge of the symmetry breaking associated with these phase transitions, by utilizing and developing new sensitive experimental probes.

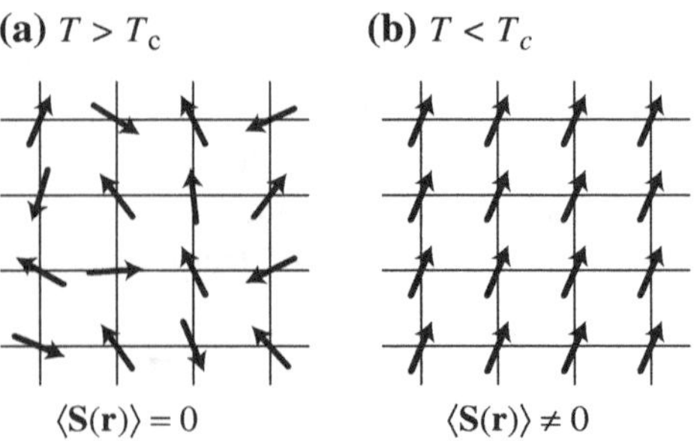

Fig. 1.1 **a** Disordered and **b** ordered spin states

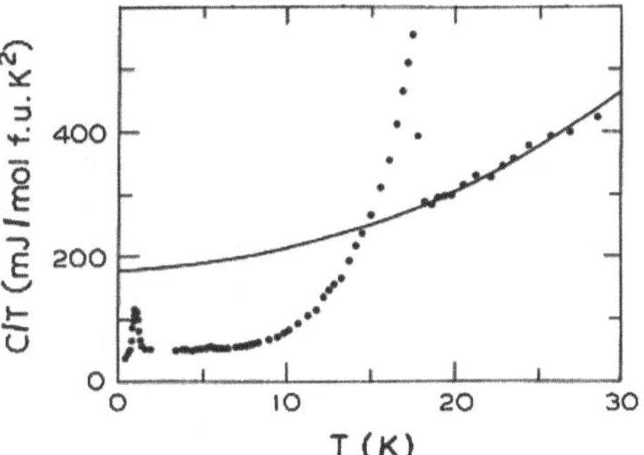

Fig. 1.2 Specific heat divided by temperature C/T as a function of temperature in URu_2Si_2. Two phase transitions at $T_0 = 17.5$ K and $T_c = 1.4$ K are clearly resolved by the C/T jumps [4]

The origin of hidden-order phase transition at T_0 and its order parameter have never been clarified. In order to solve this question, we have tried to identify what symmetry is broken in the hidden-order phase. This is the most fundamental issue to address the nature of phase transition phenomena because the order parameter is introduced to describe the ordered state with reduced symmetries. We therefore develop a magnetic torque measurement system, which is sensitive for detecting the magnetic anisotropy. It should be noted that the torque is essentially zero in an isotropic system such as materials holding the cubic symmetry. The torque measurements are then usually employed to determine the magnetic anisotropy in the anisotropic materials. Here we use this technique to measure the in-plane magnetic anisotropy in the hidden-order phase of URu_2Si_2. The URu_2Si_2 crystal possesses the tetragonal symmetry, thus the in-plane torque should be zero. However, if a new electronic or magnetic state that breaks the tetragonal symmetry emerges below T_0, the in-plane torque exhibits finite twofold oscillation. Measuring in-plane magnetic torque is therefore highly advantageous for determining whether the hidden order parameter breaks the tetragonal fourfold symmetry. In following chapter, we will firstly introduce the physical properties of URu_2Si_2, and then describe the experimental details of torque measurement, results, and discussion on the symmetry breaking in the hidden-order phase in Chap. 3.

An elucidation of the nature of superconducting state in URu_2Si_2 is also an important issue. Recent thermal conductivity and heat capacity measurements have reported an unconventional chiral superconducting state in this material [7, 8]. However, there is still a number of open questions such as its relationship with the hidden-order phase. Further experimental studies with using high-quality single crystals are also required because the superconducting state in this material is very sensitive to the sample quality. We have therefore developed the precise measurements of the lower critical fields H_{c1}, which strongly depend on the superconducting gap function. We find anisotropic H_{c1} behaviors for $\mathbf{H} \parallel c$ and $\mathbf{H} \parallel a$, consistent with the chiral d-wave superconducting state reported before. Furthermore, we newly find the anomalous kink behavior of H_{c1} below T_c, which will be discussed in terms of the rotational symmetry breaking in the hidden-order phase shown in Chap. 3. In Chap. 4, we will show the details of H_{c1} measurement, results, and discussion on the relation to the results on magnetic torque measurements.

In addition to the above two studies, we have also investigated the vortex mixed state in URu_2Si_2 by using electrical and thermal transport probes and found an intriguing phase transition phenomenon inside the superconducting phase. The phase transition we newly found in this material is a melting transition between vortex lattice and liquid, reminiscent of the melting phase transition from ice to water. In Chap. 5, we will show the results of the transport properties measured in the mixed state and discuss the origin of the melting transition.

Finally we will summarize and conclude the present study in Chap. 6.

References

1. Y. Tokura, N. Nagaosa, Science **288**, 462 (2000)
2. P. Santini, R. Lémanski, P. Erdös, Adv. Phys. **48**, 537 (1999)
3. P. Santini, S. Carretta, G. Amoretti, R. Caciuffo, N. Magnani, G.H. Lander, Rev. Mod. Phys. **81**, 807 (2009)
4. T.T.M. Palstra, A.A. Menovsky, J. van den Berg, A.J. Dirkmaat, P.H. Kes, G.J. Nieuwenhuys, J.A. Mydosh, Phys. Rev. Lett. **55**, 2727 (1985)
5. M.B. Maple, J.W. Chen, Y. Dalichaouch, T. Kohara, C. Rossel, M.S. Torikachvili, M.W. McElfresh, J.D. Thompson, Phys. Rev. Lett. **56**, 185 (1986)
6. W. Schlabitz, J. Baumann, B. Pollit, U. Rauchschwalbe, H.M. Mayer, U. Ahlheim, C.D. Bredl, Z. Phys, B **62**, 171 (1986)
7. Y. Kasahara, T. Iwasawa, H. Shishido, T. Shibauchi, K. Behnia, Y. Haga, T.D. Matsuda, Y. Onuki, M. Sigrist, Y. Matsuda, Phys. Rev. Lett. **99**, 116402 (2007)
8. K. Yano, T. Sakakibara, T. Tayama, M. Yokoyama, H. Amitsuka, Y. Homma, P. Miranovic, M. Ichioka, Y. Tsutsumi, K. Machida, Phys. Rev. Lett. **100**, 017004 (2008)

Chapter 2
Heavy-Fermion Superconductor URu_2Si_2

2.1 Overview of f-Electron Heavy-Fermion Systems

Intermetallic compounds containing $4f$- or $5f$-electron elements display a wealth of electronic and magnetic states. In $4f$-electron Ce-based heavy-fermion compounds, the Kondo lattice model is appropriate for a description of the microscopic electronic state [10, 61]. In high-temperature range, the f electron in these compounds is well localized on each Ce^{3+}($4f^1$ configuration) ion and a typical Curie-Weiss behavior described by localized spin of f electrons is observed in the temperature variation of magnetic susceptibility. At lower temperatures, however, the f electrons begin to become delocalized due to a hybridization of atomic and conduction-electron wave functions. Eventually, at sufficiently low temperatures, the f electrons become itinerant with heavy effective mass about one hundred times larger than the free electron mass. The thermodynamic and transport properties in these Ce-based heavy-fermion materials are then described in terms of the Landau Fermi-liquid theory with heavy quasiparticles. The electronic specific heat coefficient γ is given by

$$\gamma = \frac{2\pi^2 k_B^2}{3} D(\varepsilon_F), \tag{2.1}$$

where $D(\varepsilon_F)$ is the density of states at the Fermi energy ε_F. The magnetic susceptibility, which exhibits the Curie-Weiss law at high temperatures, does not show the Curie-like upturn at low temperatures because of a screening of f-electron spin by the conduction electron spin. Instead, the temperature-independent Pauli paramagnetic susceptibility χ_s is usually observed. Here χ_s is given by

$$\chi_s = 2\mu_B^2 \frac{D(\varepsilon_F)}{1 + F_0^a}, \tag{2.2}$$

where F_0^a is a dimensionless Landau parameter which parameterizes a spin-antiparallel interaction between quasiparticles. Both γ and χ_s are strongly enhanced

R. Okazaki, *Hidden Order and Exotic Superconductivity in the Heavy-Fermion Compound URu_2Si_2*, Springer Theses, DOI: 10.1007/978-4-431-54592-7_2,

from those in conventional metals owing to large $D(\varepsilon_F)$ at low temperatures. The resistivity is governed by an inelastic electron–electron scattering, leading to a quadratic temperature dependence $\rho(T) = \rho_0 + AT^2$ with a large A value, which is correlated with large γ through the Kadowaki-Woods relation [29].

In such Ce-based heavy-fermion materials, the magnetic moment of localized f electrons is quenched through the Kondo effect. On the other hand, many Ce-based materials exhibit magnetic ordering at low temperatures through the Ruderman-Kittel-Kasuya-Yosida (RKKY) interaction, which favors parallel as well as antiparallel orientation of the moments at neighboring sites. Thus, the competition between Kondo effect and RKKY interaction determines the electronic and magnetic states in Ce-based compounds at low temperatures. Doniach argued these two energy scales in the Kondo lattice system [12]. The characteristic energy scale of Kondo effect is comparable to the Kondo temperature $T_K \sim W \exp[-1/J_{cf} D_c(\varepsilon_F)]$, where W, J_{cf}, and $D_c(\varepsilon_F)$ are the conduction electron bandwidth, the exchange interaction between f and conduction electrons, and the conduction electron density of states at the Fermi energy, respectively. On the other hand, the characteristic energy scale of the RKKY interaction is given by $T_{RKKY} \sim J_{cf}^2 D_c(\varepsilon_F)$. The energy gain due to magnetic ordering follows a power low for weak hybridization whereas the Kondo temperature depends on it exponentially. Thus for small $J_{cf} D_c(\varepsilon_F)$, T_{RKKY} exceeds T_K, stabilizing a magnetically ordered state. With increasing $J_{cf} D_c(\varepsilon_F)$, however, the magnetic ordering temperature is suppressed and a Fermi liquid ground state with enhanced effective mass (heavy-fermion state) is realized above a quantum critical point (QCP).

$5f$-electron U-based heavy-fermion systems also exhibit various intriguing phenomena but the microscopic view is different from that of $4f$ Ce-based materials in which the Kondo lattice model is well applied [10]. In contrast to well-localized $4f$ electrons in Ce-based systems, more broad $5f$ wave function possesses both localized and itinerant characters. Moreover, the ionic U can adopt four different valences when combined with other elements. In fact, the f-electron valence of the U ions has been discussed controversially. Usually it takes U^{4+} ($5f^2$) or U^{3+} ($5f^3$), however it is very difficult task to distinguish these two valences in U-based metallic systems where the f electrons strongly hybridize with conduction electrons.

Recently, an importance of a duality, i.e. both itinerant and localized characters, of the $5f$ electrons in several U-based materials has been recognized. In the dual model, the itinerant electrons hybridize with the conduction states and form energy bands while the localized ones form multiplets to reduce the local Coulomb repulsion. The two subsystems interact, leading to a mass enhancement of the delocalized quasiparticles. This dual picture gives a natural description of heavy-fermion superconductivity coexisting with $5f$-derived magnetism. Table 2.1 lists U-based superconducting materials with their corresponding coexisting ordered phases. There are wide varieties of coexisting magnetic phases such as antiferromagnetism with small or large ordered moments, spin density wave, ferromagnetism and hidden order. The coexistence of magnetism and superconductivity indicates that the Cooper pairs are formed magnetically mediated interactions in these systems, although the superconducting gap structures of these materials have not yet been clarified. In particular, a

Table 2.1 Uranium-based superconductors coexisting with ordered phases

Uranium-based compounds	Coexisting order	References
UPt_3	Small-moment antiferromagnetism	[60]
UPd_2Al_3	Antiferromagnetism	[18]
UNi_2Al_3	Spin density wave	[19]
UGe_2, URhGe, UIr, UCoGe	Ferromagnetism	[1, 5, 25, 52]
URu_2Si_2	Hidden order	[35, 45, 53]

highly unusual superconducting state is expected in URu_2Si_2, since it coexists with a mysterious "hidden order", whose order parameter has not been clarified so far, as described in next session.

2.2 URu_2Si_2

2.2.1 Crystal Structure

The heavy-fermion superconductor URu_2Si_2 was discovered nearly 30 years ago [35, 45, 53]. URu_2Si_2 is crystallized in the $ThCr_2Si_2$-type structure with space group $I4/mmm$ as shown in Fig. 2.1. The lattice constants are $a = 0.41279(1)$ nm and $c = 0.95918(7)$ nm at 294 K [45]. There is no distortions or changes in symmetry between 300 and 4.2 K [28].

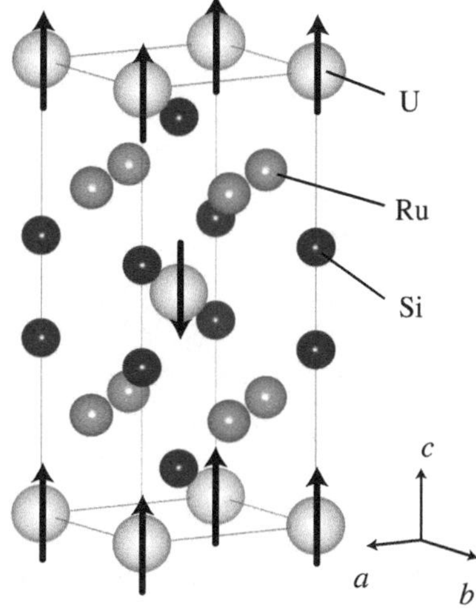

Fig. 2.1 The crystal structure of URu_2Si_2 with a body-centered tetragonal $ThCr_2Si_2$ structure. *Arrows* on the U atoms show tiny antiferromagnetic moments indicated by the neutron scattering experiments [8]

2.2.2 Thermodynamic Properties

Figure 2.2a, b show the specific heat divided by temperature C/T of URu_2Si_2 as a function of T^2 and T, respectively. At $T_0 = 17.5$ K, the specific heat shows a distinct jump, indicating a second-order phase transition. The data $C(T)$ below T_0 could be described by an expression with an energy gap Δ, $C(T) = \gamma T + \beta T^3 + a\exp(-\Delta/T)$, where a is constant. Here γT and βT^3 are the specific heat of the ungapped electrons and lattice, respectively. The third term represents the contribution from electrons involved in the formation of the hidden order. This fitting in the temperature range between 2 and 17 K yields $\Delta \simeq 110$ K. On the other hand, $C(T)$ above T_0 is well expressed as $C(T) = \gamma T + \beta T^3$ as shown by the solid line in Fig. 2.2a. At T_0, γ is reduced from 180 to 60 mJ/molK^2, indicating that a substantial portion of the Fermi surface disappears due to the gap opening at T_0. The entropy,

$$S = \int_0^{T_0} \frac{a\exp(-\Delta/T)}{T} dT, \tag{2.3}$$

that is estimated to be $\sim 0.2\, k_B \ln 2$ per formula unit, is lost in this phase transition. This entropy loss indicates a large local moment that is detectable with neutron scattering experiments if the transition at T_0 derives from a magnetic one. The second jump of C/T at $T_c = 1.4$ K shown in Fig. 2.2b indicates the superconducting transition at T_c.

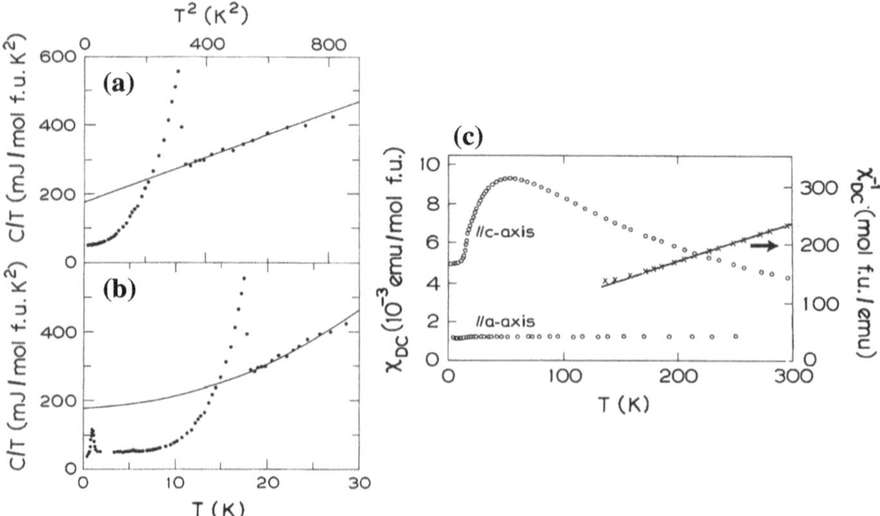

Fig. 2.2 Specific heat divided by temperature C/T as a function of **a** T^2 and **b** T. *Solid lines* show the expression $C(T) = \gamma T + \beta T^3$. **c** Temperature dependence of dc magnetic susceptibility χ_{dc} (*left axis*) and the inverse susceptibility (*right axis*) measured in a field of 2 T [45]

Figure 2.2c shows the temperature dependence of the dc magnetic susceptibility measured with $\mu_0 H = 2\,\mathrm{T}$ for $\mathbf{H} \parallel c$ and $\mathbf{H} \parallel a$ [45]. The magnetic susceptibility shows a very anisotropic behavior. In contrast to temperature-independent susceptibility for $\mathbf{H} \parallel a$, the susceptibility for $\mathbf{H} \parallel c$ shows the Curie-Weiss behavior with an effective moment $\mu_{\mathrm{eff}} = 3.51\,\mu_{\mathrm{B}}$ and a Curie-Weiss temperature $\theta_{\mathrm{CW}} = -65\,\mathrm{K}$ at high temperatures. Below 150 K, the susceptibility deviates from the Curie-Weiss law and exhibits a broad maximum around 50 K. At T_0, the susceptibility shows a small kink anomaly.

2.2.3 Transport Properties

Figure 2.3a, b display the temperature dependence of the resistivity for both current directions. In high-temperature range, the resistivity increases with lowering temperature owing to a strong incoherent scattering. The resistivity shows a maximum around 70 K, and then decreases with decreasing temperature in the lower temperature range. There, the local spin of uranium atom is disappeared as also seen in the susceptibility behavior (Fig. 2.2c), and the heavy-fermion state with the large electronic specific heat coefficient is formed. At T_0, the resistivity also exhibits anomaly as shown by the arrows in Fig. 2.3b.

Below T_0, the temperature dependence of the resistivity is well fitted by using the formula for an antiferromagnet with an energy gap Δ with an additional T^2 term from the Fermi-liquid behavior, $\rho = \rho_0 + AT^2 + BT\,(1 + 2T/\Delta)\exp(-\Delta/T)$, where ρ_0 is the residual resistivity and A and B are constant, as shown by the solid lines in Fig. 2.3b. This fitting results in $\Delta \simeq 90\,\mathrm{K}$ for $\mathbf{J} \parallel a$ and $\Delta \simeq 70\,\mathrm{K}$ for $\mathbf{J} \parallel c$, which are close to the gap value obtained from the specific heat measurement.

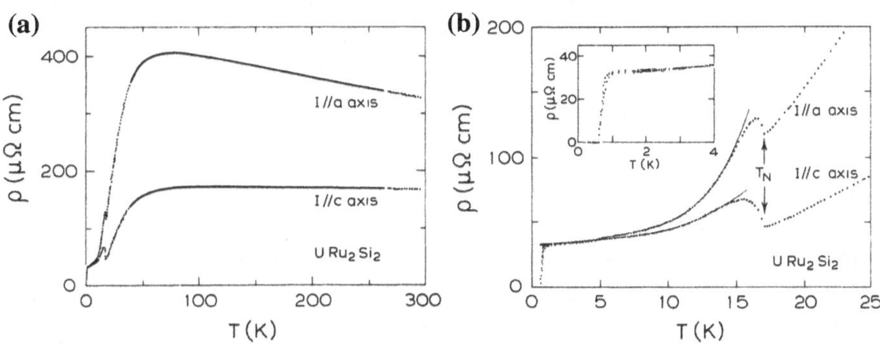

Fig. 2.3 **a** Temperature dependence of the resistivity ρ for $\mathbf{J} \parallel c$ and $\mathbf{J} \parallel a$. **b** Expanded view near the transition temperature. *Solid lines* represent the fitting using the formula appropriate for the antiferromagnet with an energy gap [46]

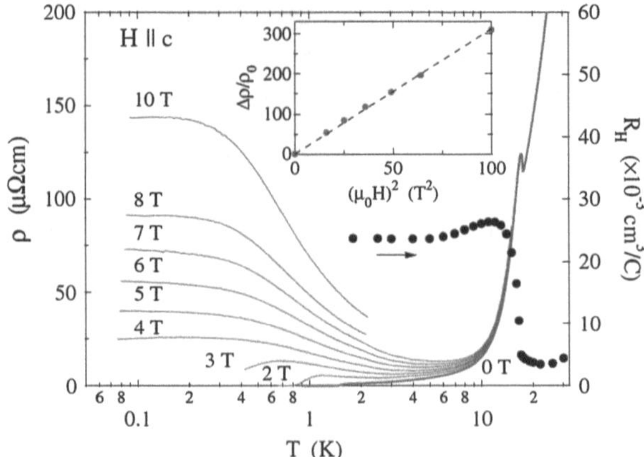

Fig. 2.4 Main panel: Temperature dependence of the resistivity ρ (*solid lines*) along the a axis in zero and finite magnetic fields ($\mathbf{H} \parallel c$) and the Hall coefficient R_H (*solid circles*). Inset: The magnetoresistance at $T \rightarrow 0$ K as a function of H^2 [30]

Figure 2.4 shows the temperature dependence of the resistivity along the a axis measured in several magnetic fields and the Hall coefficient R_H (solid circles) defined as $R_H \equiv d\rho_{xy}/dH$ at $H \rightarrow 0$ T for $\mathbf{H} \parallel c$ [30]. In zero field, the residual resistivity is extremely small $\simeq$0.5 μΩcm, yielding a large residual resistivity ratio of 670. The Hall coefficient shows a steep increase at T_0 with lowering temperature [30, 55], demonstrating a strong reduction of the carrier density in the low-temperature phase. Actually, the carrier density is estimated to be about 3×10^{26} m^{-3}, which is one order smaller than that of other heavy-fermion superconductors. This value corresponds to 0.02 carriers/U, indicating that the Fermi surfaces occupy only a small part of the Brillouin zone. The resistive behavior in magnetic fields is also remarkable; the magnetoresistance defined as $[\rho(H) - \rho(H = 0)]/\rho(H = 0)$ at $T \rightarrow 0$ K is shown as a function of H^2 in the inset of Fig. 2.4. The magnetoresistance exhibits perfect H^2 dependence and reaches about 300 at $\mu_0 H = 10$ T, indicating a carrier compensation below T_0. In such case, the magnetoresistance is given by $(\omega_c^e \tau_e)(\omega_c^h \tau_h)$, where ω_c and τ are the cyclotron frequency and the scattering time, respectively [48]. The suffixes e and h denote electron and hole, respectively. At the upper critical field H_{c2} for $\mathbf{H} \parallel c$, $(\omega_c^e \tau_e)(\omega_c^h \tau_h)$ is about 25, which is currently the highest value among the type-II superconductors.

These magneto-transport studies have revealed a highly unusual "semimetallic" electronic state below T_0, characterized by the carrier compensation with very low density. The positive Hall coefficient indicates the presence of light hole band, consistent with the results of de Haas-van Alphen (dHvA) measurements [43]. In the dHvA measurements, the branch α with the largest intensity has been resolved, which corresponds to the band 17 (hole) centered at Z point in the theoretical calculation using the relativistic linearized augmented plane wave (RLAPW) method. This hole band has a spherical shape with a light effective mass of $m_h = 13 m_0$. The dHvA measurements have also detected β and γ branches, which may correspond to the electron bands 19 and 20, respectively. The carrier compensation necessarily indicates the presence of electron Fermi surface with the same volume to the hole one. However, the volume of these electron bands is much smaller than that of main hole band. Furthermore, the magnitude of the measured electronic specific heat is much larger than the estimation taking these branches observed in the dHvA measurements [43], indicating a presence of the electron band with heavy effective mass. The observed large anisotropy of H_{c2} indicates a ellipsoidal shape of the heavy electron Fermi surface. Very recently, the quantum oscillation measurements using high-quality crystals have detected a new branch with moderate heavy mass named ω band at high magnetic fields above 25 T [6].

2.2.4 Hidden Order

Various thermodynamic and transport properties show a clear evidence of a second-order phase transition at T_0. At the early stage, this transition was regarded as an antiferromagnetic one or spin density wave (SDW) or charge density wave (CDW) transition accompanied with gap opening at the Fermi surface [35, 45]. Broholm et al. have performed elastic neutron scattering experiments and found that a weak antiferromagnetic ordering with $\mathbf{Q} = (1, 0, 0)$ appears below T_0, the magnetic structure of which is shown in Fig. 2.1 [8]. However, the magnetic moment has extremely small value of $\sim 0.03\,\mu_B$/U, in highly incompatible with a large entropy reduction ΔS associated with the transition at T_0. Moreover, neutron-scattering Bragg intensity of the (100) magnetic reflection at ambient pressure seriously differs between the samples as shown in Fig. 2.5, indicating that the weak antiferromagnetic order with $\mathbf{Q} = (1, 0, 0)$ is not intrinsic to this low-temperature phase [3].

The origin of this weak ordered moment has been revealed by several microscopic experiments including the nuclear magnetic resonance (NMR) measurements under hydrostatic pressure [37]. Figure 2.6a shows the temperature variation of the NMR spectra obtained under pressure $P = 8.3$ kbar. In contrast to a single-peak feature in the paramagnetic phase ($T = 20$ K), satellite peaks appear below T_0 around the centered peak and these develop instead of the suppression of the central peak with decreasing temperature. This result strongly indicates that the antiferromagnetic and paramagnetic phases exist in different parts of the sample volume. Figure 2.6b displays the internal field $H_{\rm in}$ and the antiferromagnetic (AF) volume fraction as a function of pressure obtained from the NMR spectra. With increasing pressure, only the AF volume fraction is varied while the internal field is almost constant. In Fig. 2.6b, the neutron Bragg scattering intensity I_B is also plotted as a function of pressure [2]. The AF volume fraction and I_B behaviors well coincide, indicating that the weak ordered moment described before originates from a tiny amount of the AF region inside the crystals.

In addition to the neutron diffraction and NMR studies, more powerful probes for detecting electronic or magnetic ordering such as the resonant x-ray scattering

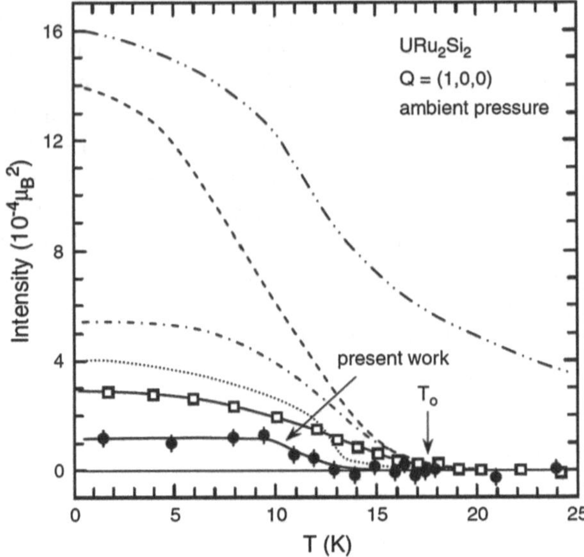

Fig. 2.5 Neutron-scattering intensity of the magnetic Bragg peak of URu_2Si_2 at $\mathbf{Q} = (1, 0, 0)$ as a function of temperature. Several experimental data reported by Amitsuka et al. (*closed circles*) [3] (*open squares*) [2], Broholm et al. (*dash-double-dotted line*) [8], Mason et al. (*broken line*) [36], Fak et al. (*dash-dotted line*) [15], and Honma et al. (*dotted line*) [24], are plotted

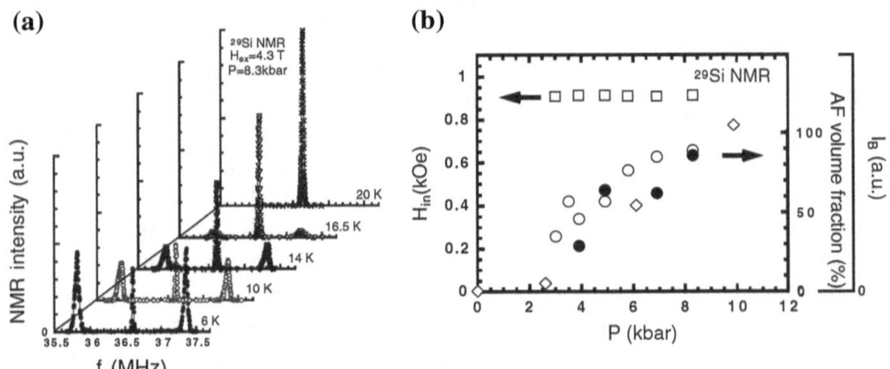

Fig. 2.6 **a** Temperature evolution of ^{29}Si NMR spectra measured under pressure P. The external magnetic field $\mu_0 H_{ex} = 4.3\,\mathrm{T}$ is applied for $\mathbf{H} \parallel c$. **b** Pressure dependence of the internal field H_{in} (*open squares*), antiferromagnetic (AF) volume fraction (*circles*), intensity of neutron Bragg scattering I_B (*open diamonds*). The data obtained by comparing the integrated intensities of the NMR resonance lines arising from the paramagnetic (PM) regions and the AF regions are plotted by the *closed circles*. The volume fraction estimated from the reduction in the values of the product of the spectral intensity and temperature is plotted by the *open circles* [37]

Table 2.2 Various theoretical models proposed for the hidden-order phase in URu_2Si_2

	References
Quadrupole	[20, 42, 50, 51]
Orbital AFM	[9]
Octupole	[32]
Hexadecapole	[23, 34]
Modulated spin liquid	[47]
Unconventional SDW	[26]
SDW and local moments	[39]
Helicity order	[62]
Dotriacontapole	[11, 27]
Spin nematic state	[16]
Hybridization wave	[13]

(RXS) [4, 63] study have been performed to identify the ordered state of this low-temperature phase. Despite much effort, however, the origin of this transition at T_0 and the order parameter in the low-temperature phase have never been unveiled, thus the term "hidden order" is adapted for the mysterious phase appearing at T_0 [40, 57]. To investigate the nature of the hidden order, a large number of microscope scenarios have been theoretically proposed. These theories can be roughly divided into three groups, in which the $5f$ electrons are considered to be localized [9, 20, 23, 32, 34, 42, 47, 50, 51], itinerant [11, 13, 14, 16, 17, 26, 27, 39, 62], or both simultaneously (dual model) [44, 58]. A portion of the proposed theories is shown in Table 2.2, but it still remains controversial. In the tetragonal crystal lattice of URu_2Si_2, ten symmetries are allowed for the local order parameter at zero field [32]. Based on this symmetry classification, several higher-order multipolar states, some of which stand on the localized $5f$-electron picture, have been proposed. Indeed, the crystal electric field calculated on the basis of the quadrupolar ordering of localized f electrons has well explained the thermodynamic properties of URu_2Si_2 [50, 51]. On the other hand, recent spectroscopic measurements show an importance of the itinerant character of f electrons in this material [7, 31, 49, 54, 64]. Nevertheless, the origin of the phase transition at T_0 in URu_2Si_2 is still puzzling and the nature of hidden-order phase has became a issue of great interest in heavy-fermion physics.

2.3 Unconventional Superconductivity in URu_2Si_2

In URu_2Si_2, the superconductivity emerges at $T_c = 1.4\,\mathrm{K}$, far below the hidden-order transition temperature, as seen in the heat capacity and resistivity measurements. The NMR measurements show that $1/T_1$ is proportional to T^3 with no coherence peak just below T_c [33] and the low-temperature specific heat measurements have revealed a non-BCS temperature dependence of C/T [21]. These experiments clearly demonstrate an unconventional superconducting state in URu_2Si_2. Furthermore, several pressure studies show an unusual temperature-pressure phase diagram as depicted in Fig. 2.7 [3]; superconductivity is suppressed with increasing pressure and disappears when the system undergoes a large-moment antiferromagnetic phase above P_M. Therefore, in URu_2Si_2, superconductivity coexists with the hidden-order phase but competes with antiferromagnetic one. This behavior is sharply contrast to other heavy-fermion systems in which superconductivity emerges under magnetically ordered phase. However, interestingly, the experimentally-observed Fermi surfaces in the hidden-order and the antiferromagnetic phases are very similar [22, 41].

Comparing to a large amount of the research on the hidden-order phase, however, the study of superconducting state of URu_2Si_2 is much less explored so far. Recent thermal conductivity measurements using a extremely clean single crystal has provided a number of valuable information to address the nature of unconventional superconducting state in URu_2Si_2 [30]. The thermal conductivity is a powerful experimental tool to measure low-lying excitations in superconductors [38]. Their experimental results are summarized in Fig. 2.8. Figure 2.8a displays the thermal conductivity divided by temperature $\kappa/T(T \to 0)$ as a function of H/H_{c2} measured for both magnetic field directions. Here, κ/T behaviors are highly non-monotonic as a function of H. Figure 2.8b shows the low-field $\kappa(H)/T$ behaviors. For both field directions, $\kappa(H)/T$ show $\sqrt{H}$ dependence with increasing H from zero field and saturate around H_s as shown by the arrow in Fig. 2.8b. This plateau behavior is reminiscent of those in the multi-band superconductors such as MgB_2 [59] and $PrOs_4Sb_{12}$ [56], indicating that a substantial portion of quasiparticles contributed from light bands is already restored at H_s, which is much below H_{c2}. In the case of URu_2Si_2, H_s is regarded as a virtual upper critical field of the light hole band. Note

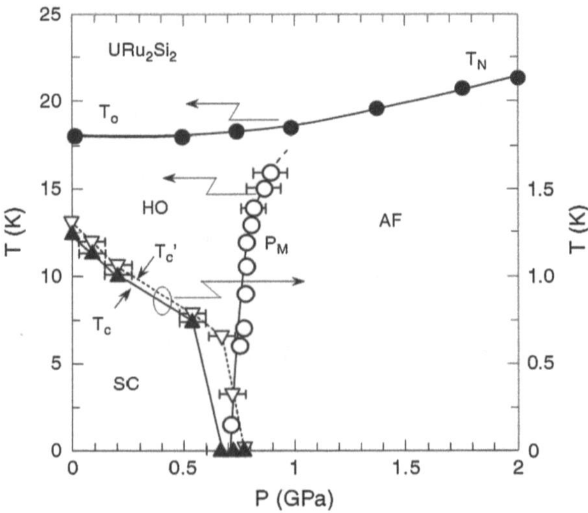

Fig. 2.7 Temperature-pressure phase diagram of URu_2Si_2 determined by ac susceptibility and elastic neutron scattering measurements [3]

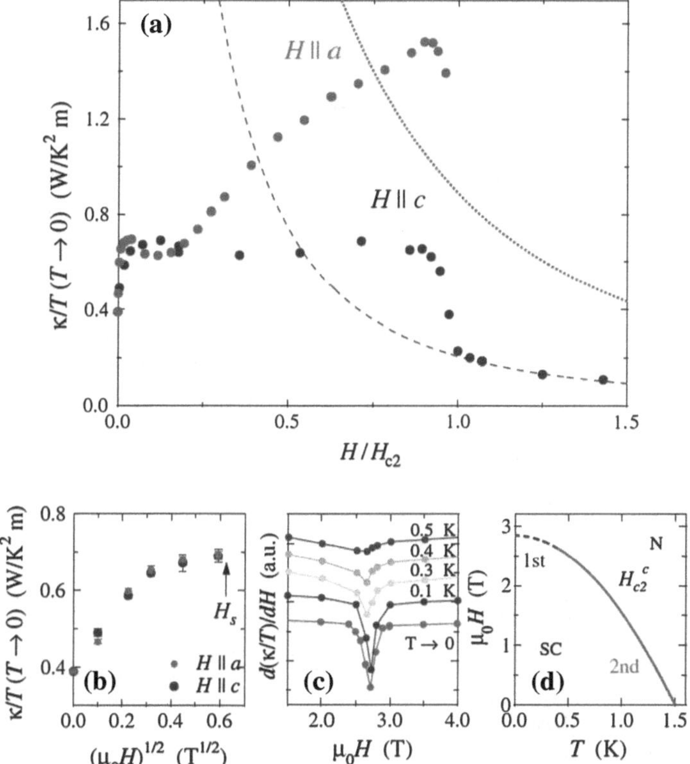

Fig. 2.8 **a** The thermal conductivity divided by temperature $\kappa/T(T \to 0)$ as a function of H/H_{c2}. The *dashed* and *dotted lines* represent the expected κ/T by the Wiedemann-Franz law for $\mathbf{H} \parallel c$ and $\mathbf{H} \parallel a$, respectively. **b** κ/T as a function of $\sqrt{H}$ at low fields. **c** The field derivative of κ/T for $\mathbf{H} \parallel c$ plotted as a function of H. **d** $H - T$ phase diagram for $\mathbf{H} \parallel c$ [30]

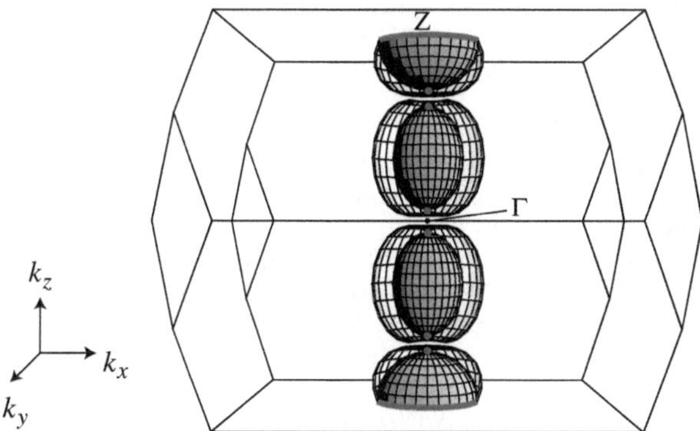

Fig. 2.9 The schematic figure of the Fermi surface (*opaque*) and superconducting gap structure (transport) inferred from thermal conductivity measurements. *Thick points* and *lines* at the Fermi surface denote the nodal part [30]

that this low-field $\sqrt{H}$ behavior indicates a presence of line nodes in the light hole band.

At high fields, on the other hand, $\kappa(H)/T$ is highly anisotropic. For $\mathbf{H} \parallel c$, $\kappa(H)/T$ is almost flat up to H_{c2}, indicating that the quasiparticle excitation is negligibly small in this field direction. On the other hand, for $\mathbf{H} \parallel a$, $\kappa(H)/T$ steeply increases with increasing H, which is interpreted as a Doppler shift effect due to nodes. Because the line node induces the quasiparticle excitation for all field directions, this anisotropic excitation behaviors above H_s strongly indicate the presence of point nodes along the c axis in the heavy electron band.

They also find a first-order nature of the phase transition at H_{c2} at low temperatures (Fig. 2.8c, d), which originates the Pauli paramagnetic pair-breaking effect expected in spin-singlet superconductors. From the symmetry analysis, they suggest a new type of unconventional superconductivity with two distinct gaps having different topology as represented by chiral d-wave which breaks time-reversal symmetry (Fig. 2.9).

References

1. T. Akazawa, H. Hidaka, T. Fujiwara, T.C. Kobayashi, E. Yamamoto, Y. Haga, R. Settai, Y. Onuki, J. Phys.: Condens. Matter **16**, L29 (2004)
2. H. Amitsuka, M. Sato, N. Metoki, M. Yokoyama, K. Kuwahara, T. Sakakibara, H. Morimoto, S. Kawarazaki, Y. Miyako, J.A. Mydosh, Phys. Rev. Lett. **83**, 5114 (1999)
3. H. Amitsuka, K. Matsuda, I. Kawasaki, K. Tenya, M. Yokoyama, C. Sekine, N. Tateiwa, T.C. Kobayashi, S. Kawarazaki, H. Yoshizawa, J. Magn. Magn. Mater. **310**, 214 (2007)
4. H. Amitsuka, T. Inami, M. Yokoyama, S. Takayama, Y. Ikeda, I. Kawasaki, Y. Homma, H. Hidaka, T. Yanagisawa, J. Phys.: Conf. Ser. **200**, 012007 (2010)
5. D. Aoki, A. Huxley, E. Ressouche, D. Braithwaite, J. Flouquet, J.P. Brison, E. Lhotel, C. Paulsen, Nature **413**, 613 (2001)
6. D. Aoki, G. Knebel, I. Sheikin, E. Hassinger, L. Malone, T.D. Matsuda, J. Flouquet, J. Phys. Soc. Jpn. **81**, 074715 (2012)
7. P. Aynajian, E.H. da Silva Neto, C.V. Parker, Y. Huang, A. Pasupathy, J. Mydosh, A. Yazdani, Proc. Natl. Acad. Sci. U.S.A. **107**, 10383 (2010)
8. C. Broholm, J.K. Kjems, W.J. Buyers, P. Matthews, T.T. Palstra, A.A. Menovsky, J.A. Mydosh, Phys. Rev. Lett. **58**, 1467 (1987)
9. P. Chandra, P. Coleman, J.A. Mydosh, V. Tripathi, Nature (London) **417**, 831 (2002)
10. D.L. Cox, A. Zawadowski, Adv. Phys. **47**, 599 (1998)
11. F. Cricchio, F. Bultmark, O. Grånäs, L. Nordström, Phys. Rev. Lett. **103**, 107202 (2009)
12. S. Doniach, Phys. B **91**, 231 (1977)
13. Y. Dubi, A.V. Balatsky, Phys. Rev. Lett. **106**, 086401 (2011)
14. S. Elgazzar, J. Rusz, M. Amft, P.M. Oppeneer, J.A. Mydosh, Nat. Mater. **8**, 337 (2009)
15. B. Fåk, C. Vettier, J. Flouquet, F. Bourdarot, S. Raymond, A. Verniere, P. Lejay, Ph Boutrouille, N.R. Bernhoeft, S.T. Bramwell, R.A. Fisher, N.E. Phillips, J. Magn. Magn. Mater. **154**, 339 (1996)
16. S. Fujimoto, Phys. Rev. Lett. **106**, 196407 (2011)
17. A.M. Gabovich, A.I. Voitenko, J.F. Annett, M. Ausloos, Supercond. Sci. Technol **14**, R1 (2001)
18. C. Geibel, C. Schank, S. Thies, H. Kitazawa, C.D. Bredl, A. Bohm, M. Rau, A. Grauel, R. Caspary, R. Helfrich, U. Ahlheim, G. Weber, F. Steglich, Z. Phys. B **84**, 1 (1991)
19. C. Geibel, S. Thies, D. Kaczorowski, A. Mehner, A. Grauel, B. Seidel, U. Ahlheim, R. Helfrich, K. Petersen, C.D. Bredl, F. Steglich, Z. Phys. B **83**, 1 (1991)

20. H. Harima, K. Miyake, J. Flouquet, J. Phys. Soc. Jpn. **79**, 033705 (2010)
21. K. Hasselbach, J.R. Kirtley, J. Flouquet, Phys. Rev. B **47**, 509 (1993)
22. E. Hassinger, G. Knebel, T.D. Matsuda, D. Aoki, V. Taufour, J. Flouquet, Phys. Rev. Lett. **105**, 216409 (2010)
23. K. Haule, G. Kotliar, Nat. Phys. **5**, 796 (2009)
24. T. Honma, Y. Haga, E. Yamamoto, N. Metoki, Y. Koike, H. Ohkuni, N. Suzuki, Y. Onuki, J. Phys. Soc. Jpn. **68**, 338 (1999)
25. N.T. Huy, A. Gasparini, D.E. de Nijs, Y. Huang, J.C.P. Klaasse, T. Gortenmulder, A. de Visser, A. Hamann, T. Gorlach, H. von Loehneysen, Phys. Rev. Lett. **99**, 067006 (2007)
26. H. Ikeda, Y. Ohashi, Phys. Rev. Lett. **81**, 3723 (1998)
27. H. Ikeda, M.-T. Suzuki, R. Arita, T. Takimoto, T. Shibauchi, Y. Matsuda, Nat. Phys. **8**, 528 (2012)
28. J.R. Jeffries, N.P. Butch, J.J. Hamlin, S.V. Sinogeikin, W.J. Evans, M.B. Maple, arXiv:1002.2245v1
29. K. Kadowaki, S. Woods, Solid State Commun. **58**, 507 (1986)
30. Y. Kasahara, T. Iwasawa, H. Shishido, T. Shibauchi, K. Behnia, Y. Haga, T.D. Matsuda, Y. Onuki, M. Sigrist, Y. Matsuda, Phys. Rev. Lett. **99**, 116402 (2007)
31. I. Kawasaki, S. Fujimori, Y. Takeda, T. Okane, A. Yasui, Y. Saitoh, H. Yamagami, Y. Haga, E. Yamamoto, Y. Onuki, Phys. Rev. B **83**, 235121 (2011)
32. A. Kiss, P. Fazekas, Phys. Rev. B **71**, 054415 (2005)
33. Y. Kohori, K. Matsuda, T. Kohara, J. Phys. Soc. Jpn. **65**, 1083 (1996)
34. H. Kusunose, H. Harima, J. Phys. Soc. Jpn. **80**, 084702 (2011)
35. M.B. Maple, J.W. Chen, Y. Dalichaouch, T. Kohara, C. Rossel, M.S. Torikachvili, M.W. McElfresh, J.D. Thompson, Phys. Rev. Lett. **56**, 185 (1986)
36. T.E. Mason, B.D. Gaulin, J.D. Garrett, Z. Tun, W.J.L. Buyers, E.D. Isaacs, Phys. Rev. Lett. **65**, 3189 (1990)
37. K. Matsuda, Y. Kohori, T. Kohara, K. Kuwahara, H. Amitsuka, Phys. Rev. Lett. **87**, 087203 (2001)
38. Y. Matsuda, K. Izawa, I. Vekhter, J. Phys.: Condens. Matter **18**, R705 (2006)
39. V.P. Mineev, M.E. Zhitomirsky, Phys. Rev. B **72**, 014432 (2005)
40. J.A. Mydosh, P.M. Oppeneer, Rev. Mod. Phys. **83**, 1301 (2011)
41. M. Nakashima, H. Ohkuni, Y. Inada, R. Settai, Y. Haga, E. Yamamoto, Y Onuki, J. Phys.: Condens. Matter **15** S2011 (2003)
42. F.J. Ohkawa, H. Shimizu, J. Phys.: Condens. Matter. **11**, L519 (1999)
43. H. Ohkuni, Y. Inada, Y. Tokiwa, K. Sakurai, R. Settai, T. Honma, Y. Haga, E. Yamamoto, Y. Onuki, H. Yamagami, S. Takahashi, T. Yanagisawa, Philos. Mag. B **79**, 1045 (1999)
44. Y. Okuno, K. Miyake, J. Phys. Soc. Jpn. **67**, 2469 (1998)
45. T.T.M. Palstra, A.A. Menovsky, J. van den Berg, A.J. Dirkmaat, P.H. Kes, G.J. Nieuwenhuys, J.A. Mydosh, Phys. Rev. Lett. **55**, 2727 (1985)
46. T.T.M. Palstra, A.A. Menovsky, J.A. Mydosh, Phys. Rev. B **33**, 6527 (1986)
47. C. Pépin, M.R. Norman, S. Burdin, A. Ferraz, Phys. Rev. Lett. **106**, 106601 (2011)
48. A.B. Pippard, *Magnetoresistance in Metals* (Cambridge University Press, Cambridge, 1989)
49. A.F. Santander-Syro, M. Klein, F.L. Boariu, A. Nuber, P. Lejay, F. Reinert, Nat. Phys. **5**, 637 (2009)
50. P. Santini, G. Amoretti, Phys. Rev. Lett. **73**, 1027 (1994)
51. P. Santini, Phys. Rev. B **57**, 5191 (1995)
52. S.S. Saxena, P. Agarwal, K. Ahilan, F.M. Grosche, R.K.W. Haselwimmer, M.J. Steiner, E. Pugh, I.R. Walker, S.R. Julian, P. Monthoux, G.G. Lonzarich, A. Huxley, I. Sheikin, D. Braithwaite, J. Flouquet, Nature **406**, 587 (2000)
53. W. Schlabitz, J. Baumann, B. Pollit, U. Rauchschwalbe, H.M. Mayer, U. Ahlheim, C.D. Bredl, Z. Phys. B **62**, 171 (1986)
54. A.R. Schmidt, M.H. Hamidian, P. Wahl, F. Meier, A.V. Balatsky, J.D. Garrett, T.J. Williams, G.M. Luke, J.C. Davis, Nature **465**, 570 (2010)
55. J. Schoenes, C. Schönenberger, J.J.M. Franse, A.A. Menovsky, Phys. Rev. B **35**, 5375 (1987)

56. G. Seyfarth, J.P. Brison, M.-A. Méasson, J. Flouquet, K. Izawa, Y. Matsuda, H. Sugawara, H. Sato, Phys. Rev. Lett. **95**, 107004 (2005)
57. N. Shah, P. Chandra, P. Coleman, J.A. Mydosh, Phys. Rev. B **61**, 564 (2000)
58. A.E. Sikkema, W.J.L. Buyers, I. Affleck, J. Gan, Phys. Rev. B **54**, 9322 (1996)
59. A.V. Sologubenko, J. Jun, S.M. Kazakov, J. Karpinski, H.R. Ott, Phys. Rev. B **66**, 014504 (2002)
60. G.R. Stewart, Z. Fisk, J.O. Willis, J.L. Smith, Phys. Rev. Lett. **52**, 679 (1984)
61. P. Thalmeier, G. Zwicknagl, in *Handbook on the Physics and Chemistry of Rare Earth*, Vol. 34, cond-mat/0312540 (2003)
62. C.M. Varma, L. Zhu, Phys. Rev. Lett. **96**, 036405 (2006)
63. H.C. Walker, R. Caciuffo, D. Aoki, F. Bourdarot, G.H. Lander, J. Flouquet, Phys. Rev. B **83**, 193102 (2011)
64. R. Yoshida, Y. Nakamura, M. Fukui, Y. Haga, E. Yamamoto, Y. Onuki, M. Okawa, S. Shin, M. Hirai, Y. Muraoka, T. Yokoya, Phys. Rev. B **82**, 205108 (2010)

Chapter 3
Magnetic Torque Study on the Hidden-Order Phase

3.1 Introduction

3.1.1 Magnetic Torque

In this chapter, we show the magnetic torque study on the hidden-order phase of URu_2Si_2. Here we briefly review the magnetic torque studies and explain the purpose of our torque measurement, which is powerful tool to determine the magnetic anisotropy. Magnetic material in a uniform field is the subject to a torque, defined as the angle derivative of the free energy F,

$$\tau = -\frac{\partial F}{\partial \alpha}\hat{\gamma} + \frac{1}{\sin\alpha}\frac{\partial F}{\partial \gamma}\hat{\alpha}, \tag{3.1}$$

where τ is the magnetic torque vector and α and γ are polar and azimuthal angles of the magnetization, respectively. Here $\hat{\alpha}$ and $\hat{\gamma}$ are unit vectors for α and γ directions, respectively. The spherical coordinate is shown in Fig. 3.1a. Let us consider the case where we rotate the magnetic field $\mathbf{H}$ in the crystalline *bc* plane with the polar angle β from the *c* axis as shown in Fig. 3.1b. In this case, $\partial F/\partial\gamma = 0$, thus the torque is given by

$$\tau = -\frac{\partial F}{\partial \alpha}\hat{\gamma}, \tag{3.2}$$

the direction of which is perpendicular to the field rotating *bc* plane. Now the free energy is expressed as,

$$F = F_0 - \mu_0 V\mathbf{M}\cdot\mathbf{H} = F_0 - \mu_0 V|\mathbf{M}||\mathbf{H}|\cos(\alpha - \beta), \tag{3.3}$$

where F_0, μ_0, $\mathbf{M}$, and V are the free energy in zero field, the permeability of vacuum, the magnetization per volume, and the sample volume, respectively. The magnetic torque is then described as

R. Okazaki, *Hidden Order and Exotic Superconductivity in the Heavy-Fermion Compound URu_2Si_2*, Springer Theses, DOI: 10.1007/978-4-431-54592-7_3,

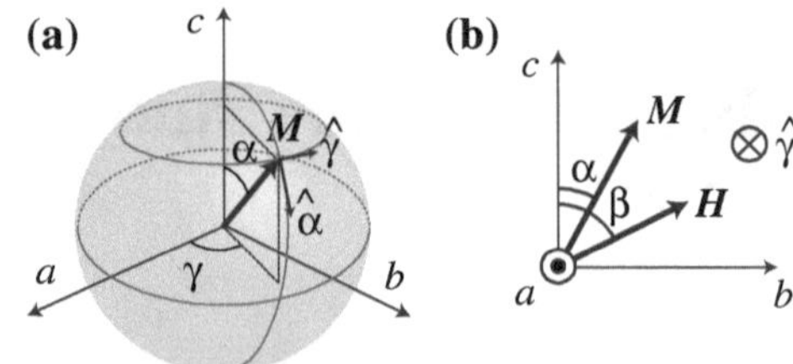

Fig. 3.1 **a** The spherical coordinate for the magnetization **M** and the magnetic field **H**. **b** The case where we rotate the field in the crystalline *bc* plane

$$\tau = \mu_0 V |\mathbf{M}||\mathbf{H}| \sin(\beta - \alpha)\hat{\gamma}, \tag{3.4}$$

and is therefore expressed as the cross product of the magnetization and the field;

$$\tau = \mu_0 \mathbf{M} V \times \mathbf{H}. \tag{3.5}$$

3.1.2 Torque in Paramagnetic Materials

Next let us consider the magnetic torque in a paramagnetic material with anisotropic magnetic susceptibility. There, the magnetization is linear to the field;

$$\mathbf{M} = \chi \mathbf{H} \tag{3.6}$$

$$= \begin{pmatrix} \chi_{aa} & \chi_{ab} & \chi_{ac} \\ \chi_{ba} & \chi_{bb} & \chi_{bc} \\ \chi_{ca} & \chi_{cb} & \chi_{cc} \end{pmatrix} \begin{pmatrix} H_a \\ H_b \\ H_c \end{pmatrix}, \tag{3.7}$$

where χ is the magnetic susceptibility tensor. Here we take a system with tetragonal or orthorhombic symmetry, where the non-diagonal elements in the susceptibility tensor are zero, and rotate the magnetic field $\mathbf{H} = H(\sin\theta, 0, \cos\theta)$ in the *ac* plane with the polar angle θ from the *c* axis (θ-rotation mode). From Eq. (3.5), we obtain the magnetic torque $\tau = (0, \tau_b, 0)$ with

$$\tau_b = \frac{1}{2}(\chi_{cc} - \chi_{aa})\mu_0 H^2 V \sin 2\theta. \tag{3.8}$$

This out-of-plane torque yields a 2-fold oscillation with respect to θ-rotation and its amplitude is proportional to the difference between *a*- and *c*-axis magnetic susceptibilities. The magnetic torque is therefore very powerful tool to investigate the magnetic anisotropy in solids. Now we show an example of the paramagnetic response of magnetic torque. Figure 3.2 shows the magnetic torque of orthorhombic $BaVSe_3$ single crystal measured in the paramagnetic phase [1]. The torque curves have been measured with rotating field in the plane containing the *c* axis, thus the torque exhibits the 2-fold oscillation whose amplitude is proportional to $\chi_{cc} - \chi_{\perp}$ as seen in Eq. (3.8),

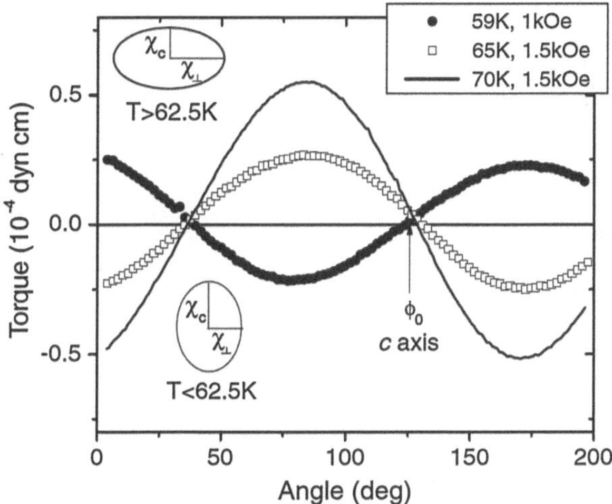

Fig. 3.2 Magnetic torque curves at several different temperatures in the paramagnetic phase of $BaVSe_3$ measured in the field-rotation plane containing the c axis. Schematic presentation of susceptibility ellipsoid in the field-rotation plane is also shown for two cases realized at different temperatures. χ_c is the susceptibility along the c axis and $\chi_\perp$ ($\approx\chi_{aa} \approx\chi_{bb}$ in this material) is the susceptibility in the direction perpendicular to the c axis in the field-rotation plane. θ_0 corresponds to the c-axis direction angle [1]

where $\chi_\perp(\approx\chi_{aa} \approx\chi_{bb}$ in this material) is the susceptibility in the direction perpendicular to the c axis in the field-rotation plane. Temperature-dependent magnetic anisotropy $\chi_{cc} - \chi_\perp$ is clearly resolved with the drastic change in the amplitude of the 2-fold torque oscillation with temperatures.

We also show the results of high-field torque measurements. Figure 3.3 depicts the magnetic torque $\tau(\theta)$ measured on the paramagnetic organic molecular conductor α-(BEDT-TTF)$_2$KHg(SeCN)$_4$ single crystal at several fields [2]. Here, the field is rotated in the ab plane with the angle θ which is taken from the a-axis direction. In this case, the torque is given by $\frac{1}{2}(\chi_{aa} - \chi_{bb})\mu_0 H^2 V \sin 2\theta$ from Eq. (3.5). As shown by the dashed curves in Fig. 3.3, $\tau(\theta)$ exhibits 2-fold oscillations. Further, $\tau(\theta)$ shows additional oscillations around $\theta = 30$ and $150°$ superimposed on the 2-fold oscillations, which derive from the quantum oscillation in χ_{bb}. The inset of Fig. 3.3 shows the H^2 dependence of the amplitude of 2-fold oscillation torque as is expected from Eq. (3.5). This H^2 dependence of magnetic torque is highly advantageous to sensitively measure the susceptibility at high magnetic fields, compared with magnetization measurements. In particular, the de Haas-van Alphen (dHvA) measurements using the torque technique have been performed extensively for organic materials [2, 3], transition-metal oxides [4], pnictides [5], and heavy-fermion systems [6].

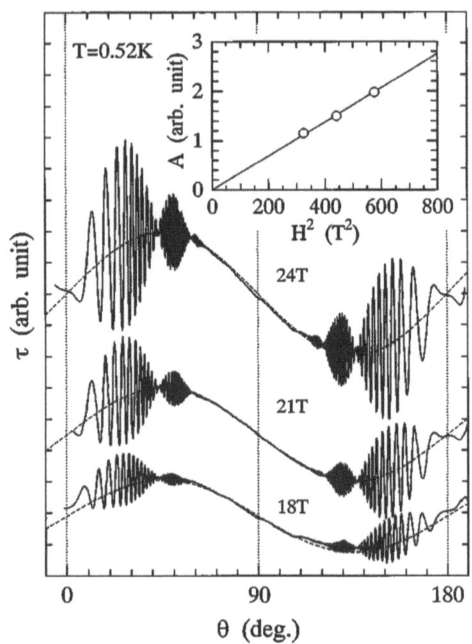

Fig. 3.3 Magnetic torque curves of the organic molecular conductor α-(BEDT-TTF)$_2$KHg(SeCN)$_4$ as a function of θ. The *dashed curves* show the $\sin 2\theta$ dependence resulting from the anisotropy of the magnetic susceptibility. The *inset* demonstrates the H^2 dependence of the torque amplitude [2]

3.1.3 Purpose of the Torque Study

In present study, we utilize the torque measurement to clarify the symmetry breaking in the hidden-order phase of URu_2Si_2 crystallizing in the tetragonal system. Here, let us consider the in-plane torque $\tau(\phi)$ with rotating field $\mathbf{H} = H(\cos\phi, \sin\phi, 0)$ in an ab plane with azimuthal angle ϕ from the a axis (ϕ-rotation mode). In this case, the magnetic torque is given by $\tau = (0, 0, \tau_c)$ with

$$\tau_c = \frac{1}{2}\mu_0 H^2 V[(\chi_{aa} - \chi_{bb})\sin 2\phi - 2\chi_{ab}\cos 2\phi]. \tag{3.9}$$

In a system with tetragonal symmetry, this in-plane torque should be zero because $\chi_{aa} = \chi_{bb}$ and $\chi_{ab} = 0$. However, finite values of the in-plane torque may appear if a new electronic or magnetic state that breaks the tetragonal symmetry emerges. In such case, rotational symmetry breaking is revealed by $\chi_{aa} \neq \chi_{bb}$ or $\chi_{ab} \neq 0$, depending on the orthorhombicity direction. Thus, such in-plane torque measurement on URu_2Si_2 provides a stringent test of whether the hidden-order parameter breaks the crystalline 4-fold symmetry. We note that this orthorhombicity may affect the out-of-plane torque shown in Eq. (3.8). However, such contribution is negligibly small compared with the leading term in Eq. (3.8) that is proportional to $\chi_{cc} - \chi_{aa}$.

3.2 Experiment

3.2.1 Samples

The single crystals of URu_2Si_2 were provided by Dr. Yoshinori Haga, Dr. Tatsuma Matsuda, Dr. Etsuji Yamamoto, and Prof. Yoshichika Onuki at Advanced Research Center, Japan Atomic Energy Agency. The single crystals were grown by the Czochralski pulling method under an argon gas atmosphere in a tetra-arc furnace, using an electro-transport-purified uranium metal as a primary material [7]. The URu_2Si_2 ingot was also subsequently annealed using the electro-transport method under ultrahigh vacuum. The annealed ingot is shown in Fig. 3.4. In URu_2Si_2, double superconducting transition due to an inhomogeneous sample quality within the crystal has often been observed in specific heat and magnetization measurements [8–10]. Figure 3.5 shows the temperature dependence of the resistivity measured with the five single crystals taken from the ingot shown in Fig. 3.4 and the inset of Fig. 3.5 (the cross-sectional view). Although the high-temperature resistive behaviors including the hidden-order phase transition temperatures well coincide with each other, the low-temperature behavior strongly depends on the position in the ingot. In particular, the superconducting transition temperature strongly depends on the residual-resistivity ratio.

Figure 3.6 shows the temperature dependence of the specific heat C/T measured with the single-crystal ingot (upper data labeled as "bulk") and a small piece cut from the ingot (lower data labeled as "No. 1"). The upper data taken from the ingot immediately after annealing indeed shows the double superconducting transition at $T \simeq 1.0$ K and $\simeq 1.4$ K, indicating that the bulk sample contains a finite amount of the lower-T_c crystals such as "No. 5" sample shown in Fig. 3.5b. However, the lower data taken from the small piece cut from the ingot exhibits a single transition jump. Note that this "No. 1" sample has very low ρ/ρ_{RT} value as shown in Fig. 3.5b. In present study, we use such high-quality URu_2Si_2 single crystals which exhibit the single superconducting transition at $T = 1.4$ K.

Figure 3.7a shows the temperature dependence of the resistivity ρ of the small piece of single-crystalline URu_2Si_2 used in present study. The resistivity was measured along the a axis (electrical current $\mathbf{J} \parallel a$) in zero magnetic field. A small dip at $T_0 = 17.5$ K indicates the hidden-order transition. As shown in the Fig. 3.7b, a well-defined superconducting transition at $T_c = 1.4$ K was observed, which is confirmed by the specific heat measurement. Below 6 K down to T_c, the resistivity depends on the temperature T as $\rho = \rho_0 + AT^2$. Our single crystal exhibits an exceptionally low

Fig. 3.4 URu_2Si_2 ingot annealed with the solid-state electro-transport method [7]

URu_2Si_2 #15

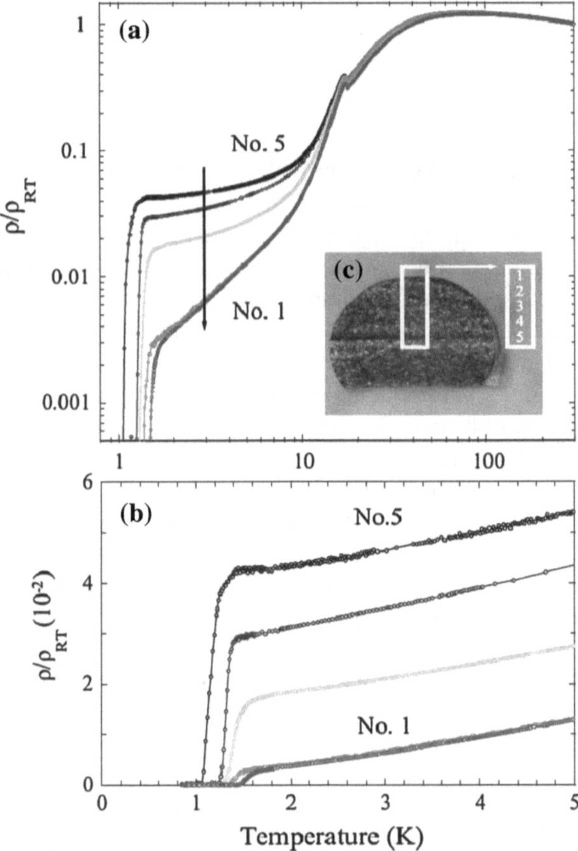

Fig. 3.5 **a** Temperature dependence of the resistivity ρ normalized by the value obtained at room temperature ρ_{RT}. The resistivity of five single crystals taken from the ingot shown in (**c**) were measured. **b** Expanded view near the superconducting transition. **c** A part of URu_2Si_2 ingot [7]

residual resistivity $\rho_0 \simeq 0.5\,\mu\Omega$cm and a large residual resistivity ratio (RRR) of 670 defined as $\rho(T = 300\,\mathrm{K})/\rho_0$, which attest the highest crystal quality currently achievable.

3.2.2 Methods

The magnetic torque is measured by a micro-tip cantilever with the flexion mode [11]. A schematic figure of the magnetic torque measurement system is shown in Fig. 3.8. We used commercially available silicon piezoresistive micro-tip cantilevers PRC120, fabricated by Seiko Instruments Inc. for the atomic force microscopy.

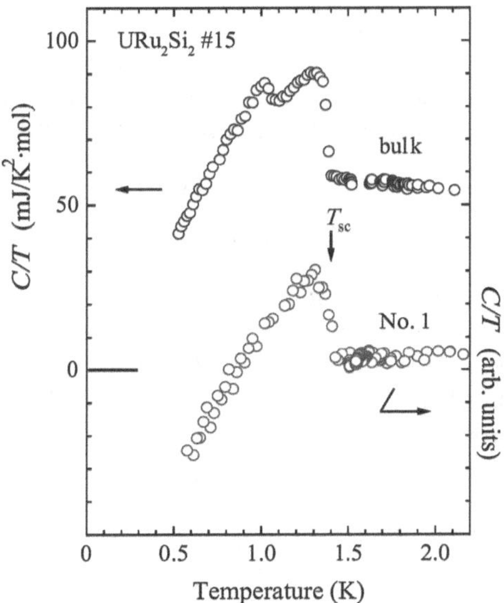

Fig. 3.6 Temperature dependence of the specific heat C/T of URu_2Si_2. *Upper* data is a specific heat of single crystal after the annealing before cutting. *Lower* one is a specific heat of single crystal after taking a small piece of the crystal by cutting the annealed ingot [7]

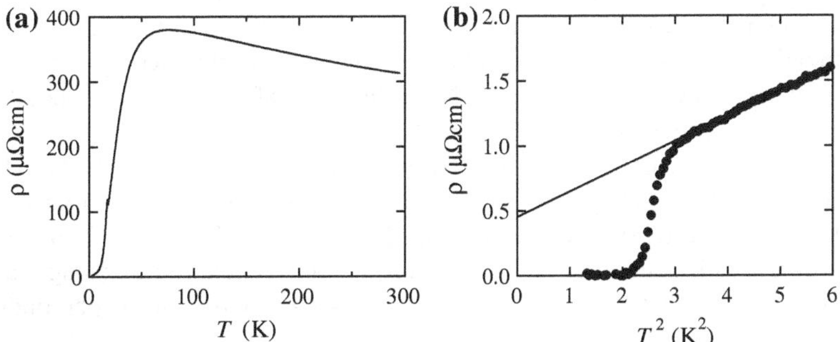

Fig. 3.7 **a** Resistivity ρ as a function of T. **b** Resistivity ρ at low temperatures as a function of T^2. The *solid line* shows a fitting by an expression $\rho = \rho_0 + AT^2$

A short cantilever for torque detection and a reference lever are incorporated on the same platform. This monolithic construction is advantageous for precise measurements, since the background signal originating from the temperature and magnetic field dependence of the cantilever can be effectively canceled out [4]. A sample is mounted at the end of a cantilever beam with a tiny amount of Araldite epoxy. We used a current source (KEITHLEY 6221A) and a voltmeter (KEITHLEY 2182A) to measure the resistivity change in the cantilever.

We construct a Wheatstone bridge circuit (Fig. 3.8) to precisely measure the resistance of the cantilever. We first make a balance of the circuit $R_s R_1 = R_R R_2$ by adjusting variable R_2 resistor, and then rotate the magnetic field as described later. Here the change of the resistance of cantilever ΔR from the balance condition is given by

$$\frac{\Delta R}{R_s} = \frac{R_{all}}{R_1 R_s} \frac{\Delta V}{I_0}, \tag{3.10}$$

where $R_{all} = R_s + \Delta R + R_R + R_1 + R_2 \simeq R_s + R_R + R_1 + R_2$ is the total resistance of the loop ($R_s \simeq 500\,\Omega$ is the resistance of the cantilever with the sample, $R_R \simeq 500\,\Omega$ is the resistance of the reference lever, R_1 and R_2 are the load resistances of $\simeq 500\,\Omega$ as shown in Fig. 3.8), ΔV is the measured voltage from the balance condition, and I_0 is the excitation current.

The magnetic torque τ is then given by [11]

$$\tau = \frac{at^2\beta}{2\pi_L} \frac{\Delta R}{R_s} \tag{3.11}$$

where a is the leg width (4 μm), t is the leg thickness (5 μm), and π_L is the piezoresistive coefficient ($4.5 \times 10^{-10}\,\mathrm{m^2/N}$). Here, β is a temperature-dependent dimensionless correction parameter, which is determined from the reference tetragonal $CeCoIn_5$ single crystals. Figure 3.9 depicts the temperature dependence of the correction parameter β. At each temperature, we measure the magnetic torque as well as the magnetization by using the superconducting quantum interference device (SQUID) magnetometer to determine β value. Note that β is almost constant about 1.2 in low-temperature range (<40 K).

In the present measurements, it is very important to rotate the magnetic field **H** with very high accuracy because a slight field misalignment produce a large effect on τ due to the anisotropy between in-plane and out-of-plane susceptibility as discussed later. In order to apply **H** for any directions, we used a mechanical rotating stage (a minimum step of 1/500°.) and a vector magnet composed of two distinct superconducting

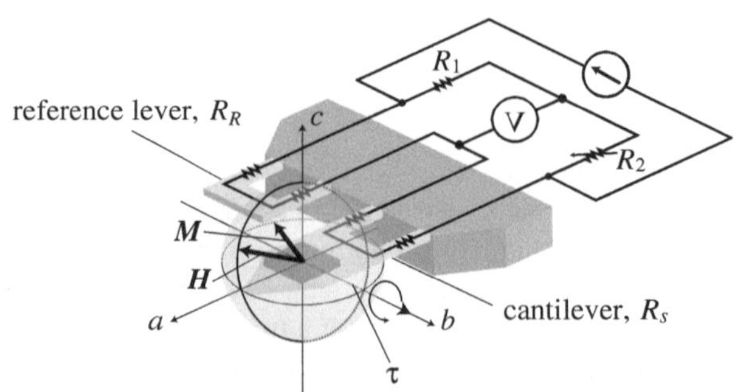

Fig. 3.8 Schematic figure of the magnetic torque measurement system using a micro-tip cantilever

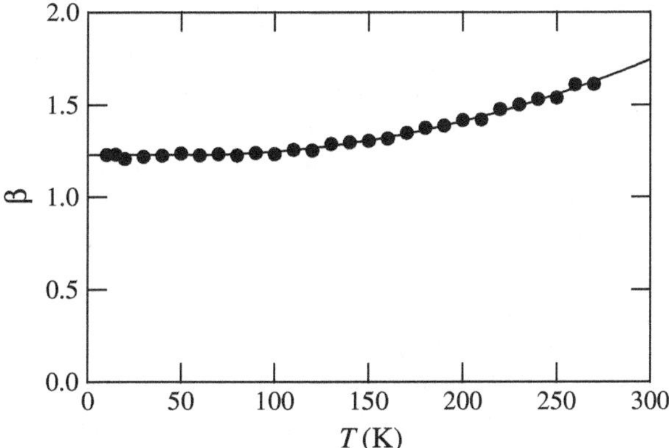

Fig. 3.9 Temperature dependence of the correction parameter β. The *solid line* is a guide for eyes. β is almost constant below $\simeq$100 K

magnets (a solenoidal superconducting magnet and a split pair superconducting magnet), which generate **H** in two mutually orthogonal directions. Figure 3.10a illustrates schematic figure of our experimental system equipped with two superconducting magnets. The solenoidal magnet generates longitudinal magnetic field $\mathbf{H}_\mathrm{L}$ and the split-pair one generates transverse field $\mathbf{H}_\mathrm{T}$ with respect to the ground. Thus, **H** is given by a vector sum of $\mathbf{H}_\mathrm{L}$ and $\mathbf{H}_\mathrm{T}$; $\mathbf{H} = \mathbf{H}_\mathrm{L} + \mathbf{H}_\mathrm{T}$. Furthermore, we utilized the mechanical rotating stage on the top of the dewar, which can rotate a cryostat with respect to the longitudinal axis. By computer controlling two magnets and the rotat-

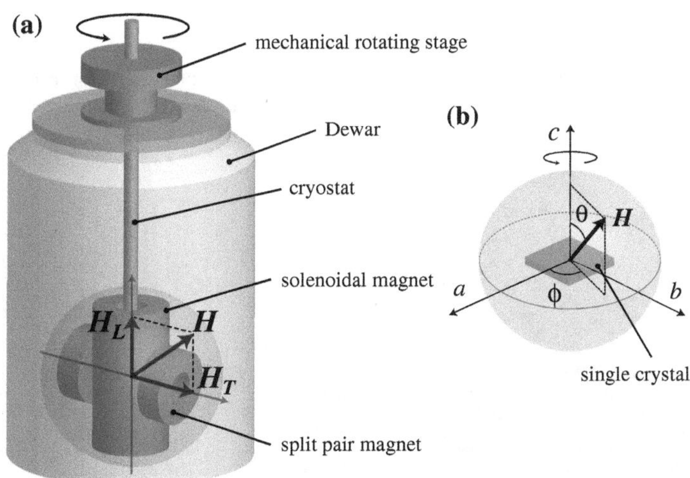

Fig. 3.10 **a** Schematic figure of our measurement system with a vector magnet and a mechanical rotating stage on the *top* of dewar. **b** Definition of the polar angle θ and the azimuthal angle ϕ

ing stage, we were able to apply **H** for various directions and control continuously within the crystalline plane with a misalignment less than 0.02°, which can be confirmed by the torque measurements in the paramagnetic phase. Figure 3.10b shows the definition of the polar angle θ and the azimuthal angle ϕ. The applied magnetic field is then described by $\mathbf{H} = |\mathbf{H}|(\sin\theta\cos\phi, \sin\theta\sin\phi, \cos\theta)$.

In our torque measurement, we employ two configurations called θ-rotation and ϕ-rotation modes. In θ-rotation mode, we put the crystal on the cantilever as the *ab* plane is parallel to the *xy* plane as shown in Fig. 3.11a and rotate the magnetic field $\mathbf{H} = H(\sin\theta, 0, \cos\theta)$ within the *ac* plane. Here we take the *xyz* coordinate axes with respect to the cantilever as drawn in Fig. 3.11a. The picture for the θ-rotation mode is shown in Fig. 3.12a. In this mode, we obtain the 2-fold-oscillation with respect to the θ rotation, the amplitude of which is proportional to $\chi_{cc} - \chi_{aa}$ (Eq. 3.8). In ϕ-rotation mode, on the other hand, we put the crystal as the *ab* plane is perpendicular to the *xy* plane as shown in Fig. 3.11b and rotate $\mathbf{H} = H(\cos\phi, \sin\phi, 0)$ within the *ab* plane. The picture is shown in Fig. 3.12b. In both rotation modes, τ is parallel to the *y* axis and works to flex the lever.

Figure 3.13 shows the experimental diagram of our torque measurement system. The components of measurement system are controlled by a personal computer through the GPIB interfaces. Experiments have beenautomatically performed

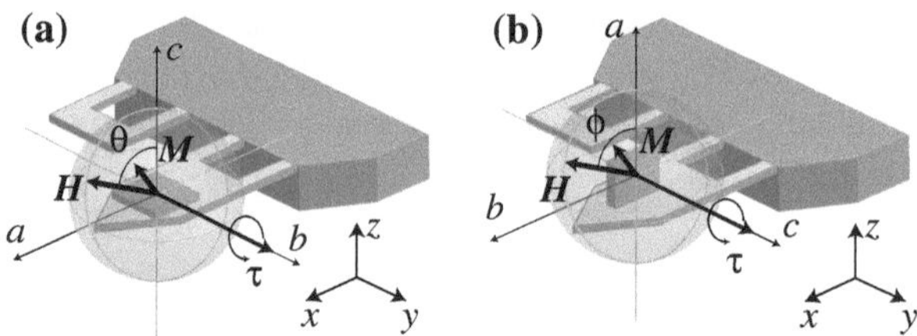

Fig. 3.11 Schematic figure of the magnetic torque measurement system for **a** θ-rotation and **b** ϕ-rotation modes. The *xyz* coordinate axes are taken with respect to the cantilever

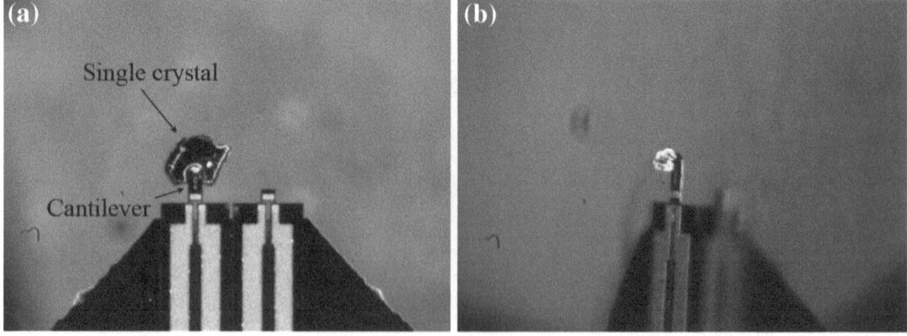

Fig. 3.12 URu_2Si_2 single crystal mounted on the micro-tip cantilever for **a** θ-rotation and **b** ϕ-rotation modes

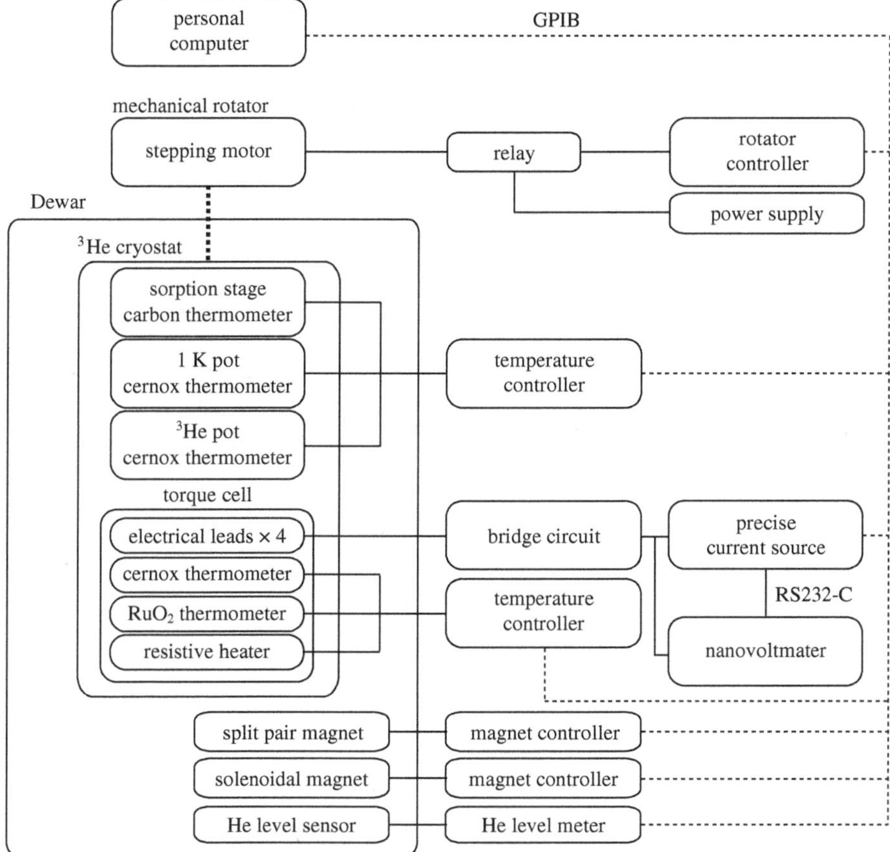

Fig. 3.13 Experimental diagram of the torque measurement

with using LabVIEW (National Instruments Inc.) programs. Figure 3.14 illustrates a schematic figure of our torque measurement system. The liquid helium dewar equipped with a vector magnet and a mechanical rotator is placed within a pit. The vacuum lines are connected with vacuum pumps through anti-vibration blocks to reduce the noise due to mechanical vibrations from the pumps.

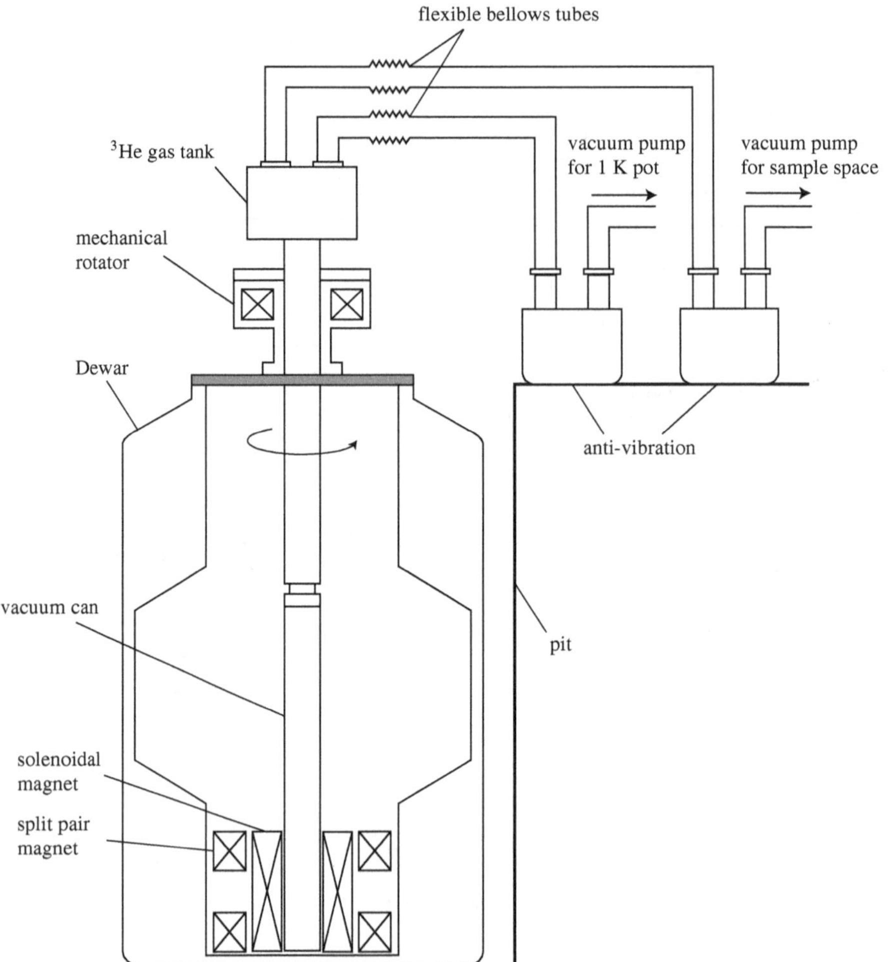

Fig. 3.14 Schematic figure of the torque measurement system

3.3 Results

3.3.1 Out-of-Plane Torque

We first show the results on the θ-rotation mode depicted in Fig. 3.11a. In this geometry, we measure the difference between a- and c-axis susceptibilities, which yields a 2-fold oscillation term with respect to θ-rotation as described in Eq. (3.8). Figure 3.15 depicts the magnetic torque as a function of θ measured on the URu_2Si_2 single crystal #8 at several temperatures. The absolute value of magnetic field is fixed at $\mu_0|\mathbf{H}| = 4\,\mathrm{T}$. Both below and above T_0, the curves are perfectly sinusoidal and can

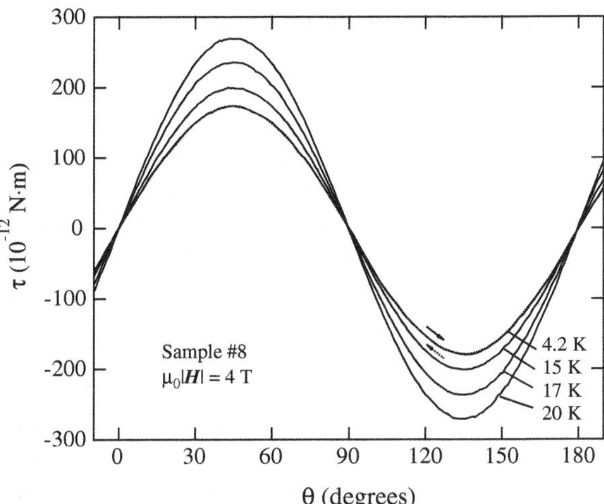

Fig. 3.15 Magnetic torque τ as a function of the polar angle θ measured on the URu_2Si_2 single crystal #8 at several temperatures. The absolute value of magnetic field is fixed at $\mu_0|\mathbf{H}| = 4\,\mathrm{T}$. Torque curves measured by rotating **H** in clockwise (*dotted line*) and anticlockwise (*solid line*) directions coincide

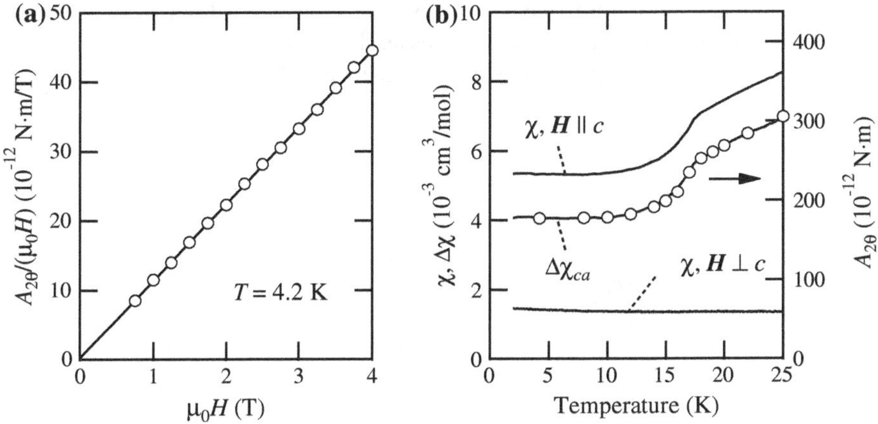

Fig. 3.16 **a** Magnetic field dependence of the amplitude of the 2-fold oscillation of the torque divided by the field, $A_{2\theta}/\mu_0 H$, at $T = 4.2\,\mathrm{K}$ for sample #8. The *solid line* is a result for linear fit. **b** *Left axis*: Temperature dependence of the susceptibility χ for $\boldsymbol{H} \perp c$ and $\boldsymbol{H} \parallel c$. Their difference $\Delta\chi_{ca} \equiv \chi_{cc} - \chi_{aa}$ is also shown. *Right axis*: Temperature dependence of $A_{2\theta}$ (*circles*)

be fitted with $\tau(T, H, \theta) = A_{2\theta}(T, H)\sin 2\theta$, where $A_{2\theta}$ is the amplitude of 2-fold oscillation. For 4.2-K data, we show the torque curves measured by rotating **H** in clockwise (dotted line) and anticlockwise (solid line) directions, which well coincide with each other.

Figure 3.16a shows the magnetic field dependence of $A_{2\theta}/\mu_0 H$ at $T = 4.2$ K. The perfect H-linear dependence of $A_{2\theta}/\mu_0 H$ with a negligible y-intercept indicates the field independent magnetic susceptibility, i.e. purely paramagnetic response, clearly showing the absence of ferromagnetic impurities as discussed later. To check the consistency of the data, we also measured the magnetic susceptibility of a different single crystal of URu_2Si_2 in the same batch using SQUID magnetometer. In Fig. 3.16b, we show the magnetic susceptibilities for $\mathbf{H} \perp c$ and $\mathbf{H} \parallel c$ and their difference $\Delta\chi_{ca} \equiv \chi_{cc} - \chi_{aa}$ as solid lines. The amplitude of $\chi_c - \chi_a$ obtained in the two types of measurements quantitatively coincide both in the hidden-order state and in the paramagnetic state above T_0.

Here we mention the impurities in URu_2Si_2 single crystals. It has been reported by several groups [12, 13] that URu_2Si_2 single crystals may contain the ferromagnetic impurities. Figure 3.17 displays the temperature variation of the magnetic susceptibility (M/H) of URu_2Si_2 measured at 2.5 Oe (open circles) and 5 kOe (closed circles) [12]. Triangles are the spontaneous magnetization M_s at 2.5 Oe obtained from the deference between these two quantities. This sample exhibits three ferromagnetic-like transitions at $T_1^* \sim 120$ K, $T_2^* \sim 40$ K, and $T_3^* \sim 16$ K as indicated by arrows in Fig. 3.17. Some of our crystals also show the sign of the ferromagnetic impurities. Figure 3.18 displays the angular variation of $\tau(\theta)$ in $\mathbf{H}$ rotating within the ac plane measured on single crystal #1. The 2-fold oscillation is due to the ac anisotropy of URu_2Si_2 as discussed before. We measured $\tau(\theta)$ by rotating $\mathbf{H}$ clockwise (dotted line) and anticlockwise (solid line) directions as shown by the two lines with arrows. In contrast to the results on sample #8 shown in Fig. 3.15, sample #1 shows the

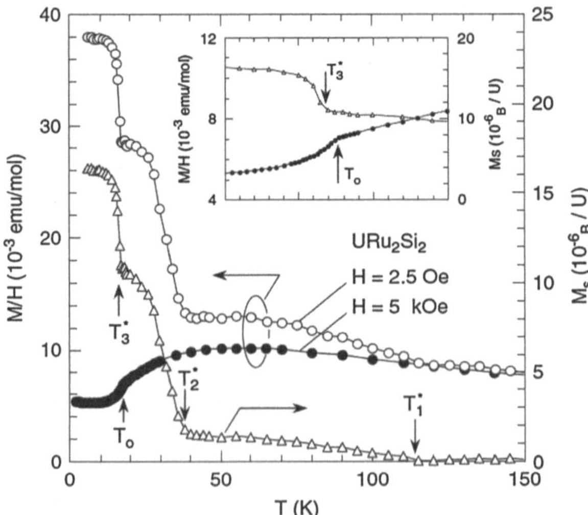

Fig. 3.17 Temperature dependence of the magnetic susceptibility of URu_2Si_2 measured at 2.5 Oe (*open circles*) and 5 kOe (*closed circles*). *Triangles* show the difference between these two quantities. The *inset* shows an enlarged view around T_0 [12]

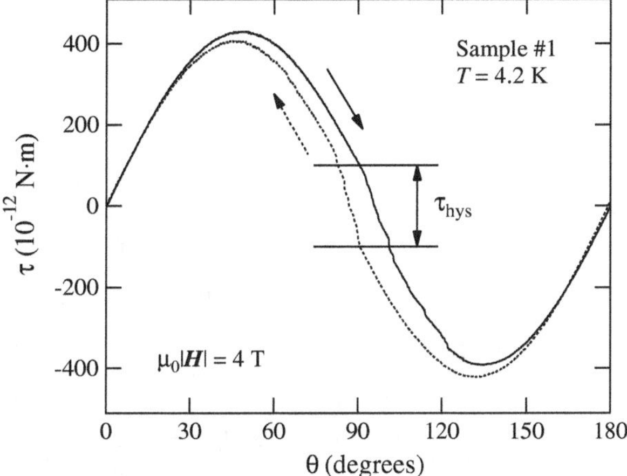

Fig. 3.18 Magnetic torque τ as a function of the polar angle θ measured on the URu_2Si_2 single crystal #1 at $T = 4.2\,K$ and $\mu_0|\mathbf{H}| = 4\,T$. Torque curves measured by rotating $\mathbf{H}$ in clockwise and anticlockwise directions are shown by *dotted* and *solid lines*, respectively

hysteresis due to the ferromagnetic impurities in the torque curves at $T = 4.2\,K$. The width of this hysteresis τ_{hys} has a peak at $\theta = 90°$, implying that this ferromagnetic impurity phase likely originates from stacking fault defects parallel to the *ab* planes.

Figure 3.19 displays the temperature dependence of the hysteresis width τ_{hys} at $\theta = 90°$ (shown in Fig. 3.18) normalized by the amplitude of the 2-fold oscillation $A_{2\theta}$ at $T = 20\,K$. In samples #1–#4, the ferromagnetic phase clearly appears below $T_C \sim 16\,K$, which is consistent with the previous report [12]. In contrast, in samples #5 and #6 the hysteresis at low temperatures is very small, and samples #7 and #8 show no sign of the ferromagnetic phase within the experimental resolution of $\sim 10^{-14}\,Nm$ (0.01 % of $|A_{2\theta}|$). We therefore use samples #7 and #8 for the ϕ-rotation measurements.

We have also measured the magnetic torque in the superconducting phase of URu_2Si_2 below $T_c = 1.4\,K$. The magnetic torque measured in superconducting phase gives several valuable information on the superconducting parameters such as the mass anisotropy and the penetration depth [14, 15]. The multiband nature of superconductors has also been investigated by the torque measurements [16]. However, the torque curves measured in the superconducting phase of URu_2Si_2 were not reversible owing to a vortex pinning, exhibiting large hysteresis around $\theta = 90°$. Such an irreversible component in torque curves makes it impossible to determine exact reversible contributions, thus we have explored the torque measurements only in the hidden-order phase in which we have obtained the completely reversible torque curves. A vortex shaking technique to reduce such irreversible components is required for future torque study in superconducting phase of URu_2Si_2.

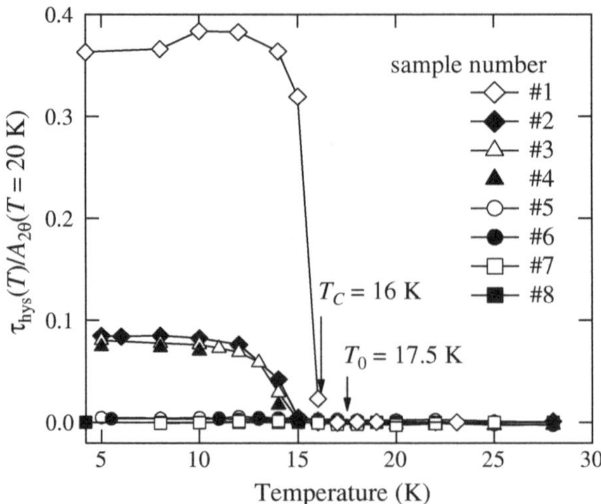

Fig. 3.19 Temperature dependence of the hysteresis width τ_{hys} at $\theta = 90°$ normalized by the amplitude of the 2-fold oscillation $A_{2\theta}$ at $T = 20\,\text{K}$

3.3.2 In-Plane Torque

We then show the results on the ϕ-rotation mode depicted in Fig. 3.11b. Figure 3.20 shows the magnetic torque as a function of the azimuthal angle ϕ measured on sample #8 in the paramagnetic phase ($T = 20\,\text{K}$, $\mu_0|\mathbf{H}| = 4\,\text{T}$). To exclude 2-fold oscillations appearing as a result of misalignment, we precisely apply **H** in the *ab* plane by controlling two superconducting magnets and a rotating stage. As seen in Fig. 3.20, the residual 2-fold component in $\tau(\phi)$ is less than $\sim 1 \times 10^{-13}\,\text{N} \cdot \text{m}$, which corresponds to a misalignment error of less than 0.02° from the *ab* plane. We note that the observed 4-fold oscillation (and higher-order terms) arises primarily from the nonlinear susceptibilities [17], which can be naturally understood from the 4-fold rotational symmetry in the tetragonal lattice of URu_2Si_2.

In Fig. 3.21a–e, we depict the temperature evolution of the in-plane torque $\tau(\phi)$ at $\mu_0|\mathbf{H}| = 4\,\text{T}$ in the paramagnetic and the hidden-order phases. The upper panels show raw torque curves as a function of the azimuthal angle ϕ at several temperatures. One can find that the torque curves shown by the upper panels apparently become asymmetric with respect to 90° rotations below T_0. Note that all torque curves are perfectly reversible with respect to the field rotation direction as discussed later. Here we employ the Fourier analysis to the raw torque curves: $\tau(\phi)$ can be decomposed as $\tau = \tau_{2\phi} + \tau_{4\phi} + \tau_{6\phi} + \cdots$, where $\tau_{2n\phi} = A_{2n\phi} \sin 2n(\phi - \phi_0)$ is a term with $2n$-fold symmetry with $n = 1, 2, \cdots$. In the middle and lower panels of Fig. 3.21a–e, the 2-fold and 4-fold components obtained from the Fourier analysis are displayed. We find that the 2-fold component $\tau_{2\phi}$, which should be zero in a crystal with tetragonal symmetry, emerges at low temperatures. The presence of the 2-fold oscillation, which follows the functional form $\tau_{2\phi} = A_{2\phi}\cos 2\phi$, indicates that $\chi_{ab} \neq 0$, whereas $\chi_{aa} = \chi_{bb}$.

Figure 3.22 depicts the temperature dependence of the 2-fold oscillation amplitude normalized by the volume $|A_{2\phi}|/V$ measured at two single crystals (#7 and #8) with different sample volumes. The inset shows the expanded view for sample #8 near the hidden-order transition. This 2-fold amplitude becomes nonzero below T_0, indicating that nonzero χ_{ab} appears in the hidden-order phase. From Eq. (3.9), we calculate $|\chi_{ab}|$ as

$$|\chi_{ab}| = \frac{1}{\mu_0 H^2}\frac{|A_{2\phi}|}{V}. \tag{3.12}$$

The calculated in-plane anisotropy $2|\chi_{ab}|/\chi_{aa}(T = 20\,\mathrm{K})$ for samples #7 and #8 is shown in the right axis of Fig. 3.22. We have also measured the magnetic susceptibility of the large single crystal (sample #0) for $\mathbf{H} \parallel [110]$ and $\mathbf{H} \parallel [1\bar{1}0]$ by using SQUID magnetometer, and evaluate $|\chi_{ab}|$ as

$$|\chi_{ab}| = \frac{1}{2}|\chi[110] - \chi[1\bar{1}0]|, \tag{3.13}$$

where $\chi[110]$ and $\chi[1\bar{1}0]$ are the magnetic susceptibilities for $\mathbf{H} \parallel [110]$ and $\mathbf{H} \parallel [1\bar{1}0]$, respectively. This SQUID data for the large crystal is also shown in Fig. 3.22 in the same scale. The in-plane anisotropy value at low-temperatures is about 10 % for the small crystal (sample #8), while it is almost zero within our experimental resolution for the large crystal (sample #0). The origin of different in-plane anisotropy values among these three samples will be discussed later.

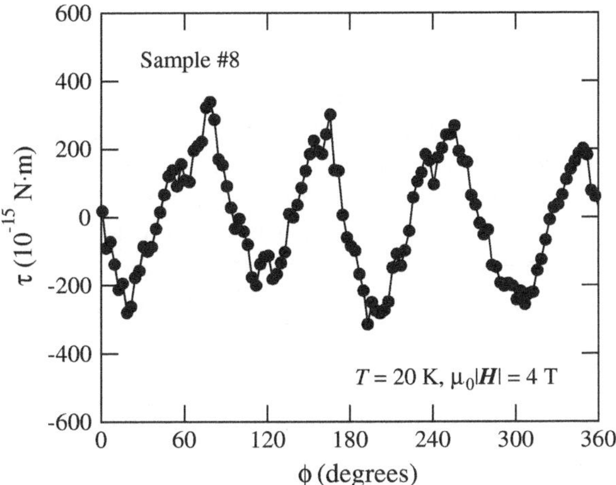

Fig. 3.20 Magnetic torque τ as a function of the azimuthal angle ϕ measured on the URu_2Si_2 single crystal #8 at $T = 20\,\mathrm{K}$ and $\mu_0|\mathbf{H}| = 4\,\mathrm{T}$

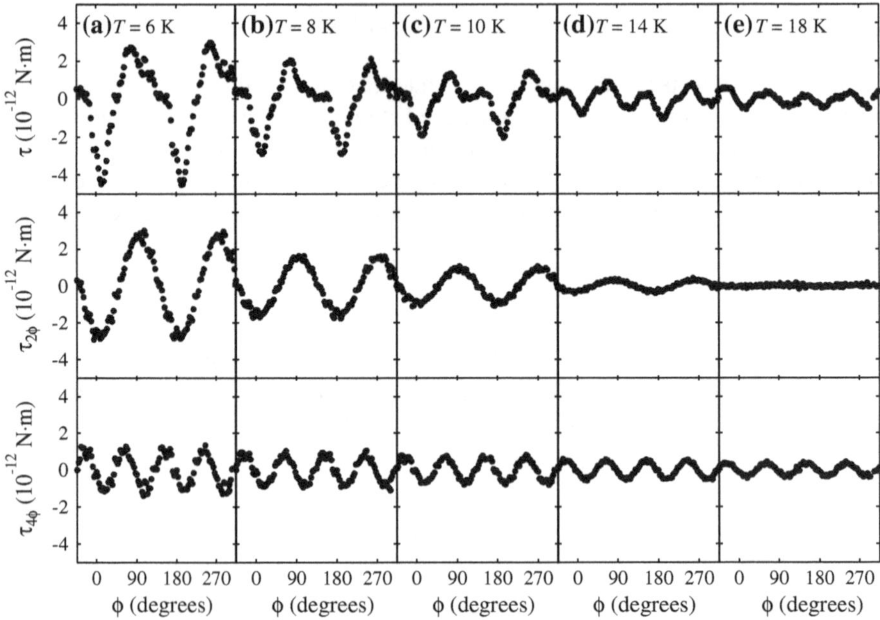

Fig. 3.21 In-plane magnetic torque measured on sample #8 at $\mu_0|\mathbf{H}| = 4\,\mathrm{T}$. *Upper panels* show raw torque curves $\tau(\phi)$ as a function of the azimuthal angle ϕ at several temperatures. *Middle* and *lower panels* show the 2-fold $\cos 2\phi$ and 4-fold $\sin 4\phi$ components of the torque curves, respectively, obtained from Fourier analysis of the raw torque curves in the *upper panels*

3.4 Discussion

We have observed the finite 2-fold oscillation component in the in-plane torque below T_0, that indicates the broken rotational symmetry in the hidden-order phase of URu_2Si_2. In this section, we will discuss the origin for observed the 2-fold oscillation including the extrinsic effects. Here we have confirmed that there is no background signals for $\mu_0|\mathbf{H}| = 4\,\mathrm{T}$ at low temperatures by the blank measurement only with Araldite epoxy. Then the 2-fold components observed below T_0 come from the crystal. As discussed below, we can exclude extrinsic effects as primary origins of the observed 2-fold oscillations in $\tau(\phi)$ in the hidden-order phase.

3.4.1 Extrinsic Effects

3.4.1.1 Effects of Misalignment

The change in the *ac* anisotropy below T_0 may contribute the 2-fold oscillations because of the tiny misalignment of the sample configuration. However, the

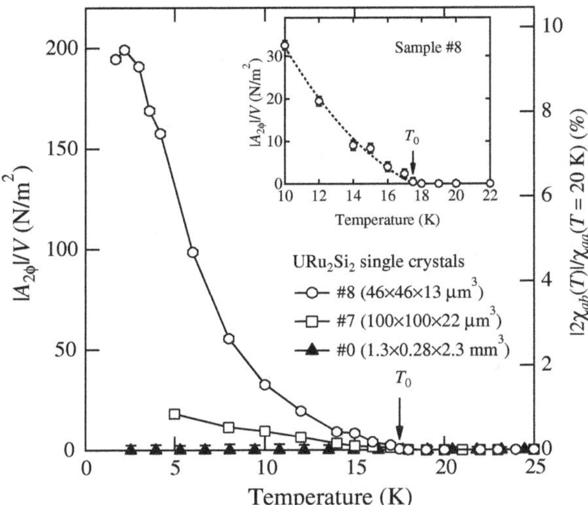

Fig. 3.22 Temperature dependence of 2-fold oscillation amplitude divided by the sample volume $|A_{2\phi}|/V$ measured for sample #7 (*open squares*) and sample #8 (*open circles*). The normalized in-plane susceptibility anisotropy $|2\chi_{ab}(T)|/\chi_{aa}(T = 20\,\text{K}) = (\chi[110] - \chi[1\bar{1}0])/\chi[100]$ is evaluated in the *right axis*. SQUID data for the large crystal (sample #0) are also shown (*triangles*) in the same scale

temperature dependence of $|A_{2\phi}|/V$ in Fig. 3.22 is clearly different from that of $\Delta\chi_{ca}$ shown in Fig. 3.16b. If this is the main source of the observed 2-fold oscillations, then the $|A_{2\phi}|/V$ should saturate at low temperatures below ~10 K, which is obviously not the case. So we can clearly exclude this possibility.

Also, there are possible thermal contraction effects in the low-temperature cryostat that could cause the temperature dependence of the misalignment. However, the thermal contraction generally becomes smaller at lower temperatures so the experimental fact that we find no 2-fold oscillations in a temperature range above T_0 up to at least 40 K indicates that such effect is irrelevant.

3.4.1.2 Effects of Sample Shape

In a magnetic field **H**, magnetic sample with a finite dimension is affected by the demagnetizing effect. The effective magnetic field $\mathbf{H}_{\text{eff}}$ for the sample is given by

$$\mathbf{H}_{\text{eff}} = \mathbf{H} - N\mathbf{M}, \tag{3.14}$$

where N is the demagnetizing tensor. Then the in-plane magnetic torque which is induced by the demagnetizing effect τ_{demag} in a crystal with tetragonal symmetry is given by

$$\tau_{\text{demag}} = \frac{1}{2}\mu_0 H^2 V \chi_{aa}^2 (N_{aa} - N_{bb}) \sin 2\phi, \tag{3.15}$$

which exhibits a 2-fold oscillation if the sample shape is anisotropic ($N_{aa} \neq N_{bb}$). However, our samples have almost plate-like shape with the thinnest direction along the c axis, then $N_{aa} \approx N_{bb} \ll 1$. Furthermore, the observed amplitude of in-plane torque is $|A_{2\phi}|/(\mu_0 H^2 V) = |\chi_{ab}| \simeq 0.05\chi_{aa}$ at low temperatures (Fig. 3.22). If this observed 2-fold oscillation originates from the demagnetizing effect, $|N_{aa} - N_{bb}|$ must be $0.1\chi_{aa}^{-1} \gg 1$, which is obviously impossible. It should also be noted that such effect should not produce strongly temperature dependent 2-fold oscillations as we observed.

3.4.1.3 Effects of Magnetic Impurities

In the samples selected for the in-plane rotation measurements, we have no detectable signals from the ferromagnetic impurities as shown in the Sect. 3.4.1.2. We show the in-plane rotation results of the $\tau(\phi)$ curves measured in several conditions at 4.2 K in Fig. 3.23. The magnetic field of 4 T is applied in the paramagnetic phase at $T > T_0$, after which we cool down the sample (field cooling condition). The torque curves are almost reproducible and show no hysteresis between the two field rotation directions. These results also confirm the null effect of ferromagnetic impurities. We add that the 2-fold oscillations onset at $T_0 = 17.5$ K as shown in the inset of Fig. 3.22, which is higher than the ferromagnetic temperature of $T_C < 16$ K, strongly suggesting that the observed 2-fold oscillations are not related to the ferromagnetic impurities.

We also note that the tiny antiferromagnetic moment of $0.03\mu_B$ [18], which is not regarded as an intrinsic property of the hidden order [12, 19, 20], cannot be an

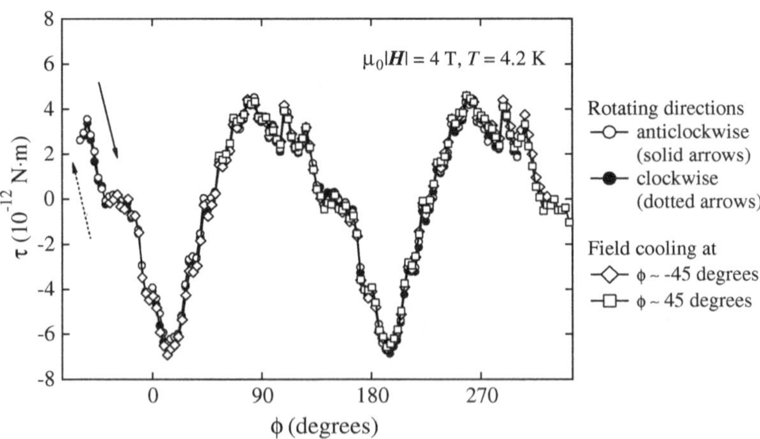

Fig. 3.23 Torque curves in different conditions. $\tau(\phi)$ measured in sample #8 with anticlockwise (*open circles*) and clockwise (*closed circles*) directions for $\mu_0|\mathbf{H}| = 4$ T at $T = 4.2$ K. Torque curves after field cooling at $\phi \sim -45°$ (*open diamonds*) and $\phi \sim 45°$ (*open squares*) are also shown

origin of this 2-fold oscillation in the plane, as spin aligns parallel to the c axis in its magnetic structure $\mathbf{Q} = (0, 0, 1)$ [equivalent to (1,0,0)].

3.4.2 Broken Rotational Symmetry in the Hidden-Order Phase

In above sections, we have excluded the extrinsic effects for the observed 2-fold oscillation in the hidden-order phase. Based on these results, we conclude that the amplitude of 2-fold oscillations is a manifestation of intrinsic in-plane anisotropy of the susceptibility, i.e.

$$\chi[110] = \chi_{aa} + \chi_{ab} \neq \chi[1\bar{1}0] = \chi_{aa} - \chi_{ab}. \tag{3.16}$$

The in-plane anisotropy that sets in precisely at T_0 indicates that the rotational symmetry is broken in the hidden-order phase. Figure 3.24 schematically show the results of our magnetic torque study. In the paramagnetic phase of URu_2Si_2, the 4-fold rotational symmetry is preserved ($\chi_{aa} = \chi_{bb}$ and $\chi_{ab} = 0$), as shown in Fig. 3.24b. In the hidden-order phase, on the other hand, the 4-fold rotational symmetry is broken ($\chi_{aa} = \chi_{bb}$ but $\chi_{ab} \neq 0$).

An intriguing question is that why has such an in-plane magnetic anisotropy not been reported before. Here we have measured several samples with different sizes as shown in Fig. 3.22. In mm-sized crystals, we observed no difference between $\chi(T)$ for $\mathbf{H} \parallel [110]$ and $\chi(T)$ for $\mathbf{H} \parallel [1\bar{1}0]$, but in samples with a smaller volume V, a nonzero $2\chi_{ab} \propto |A_{2\phi}|/V$ appeared. This may imply that the hidden-order phase forms domains with different preferred directions in the ab plane, which may be a natural consequence of the tetragonal crystal structure. A domain size on the order of tens of micrometers would explain both our results and the difficulties in observing this effect. The fact that the torque curves remain unchanged for field-cooling conditions at different field angles (Fig. 3.23) implies that the formation of

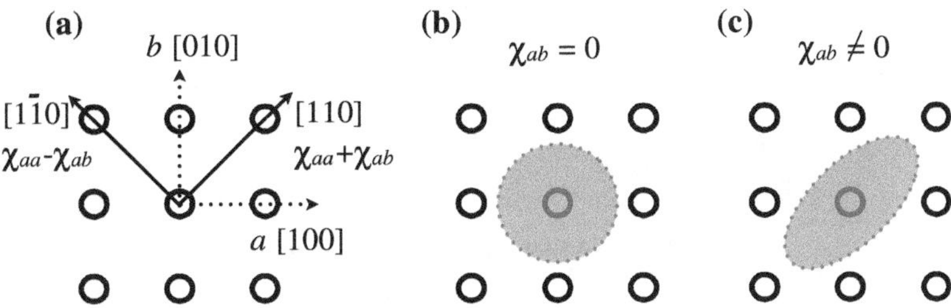

Fig. 3.24 **a** Uranium atom (*open circles*) arrangement in the ab plane of URu_2Si_2. The relevant axes and the susceptibility components for the $\chi_{aa} = \chi_{bb}$ case are also shown. Rotational symmetries in **b** the paramagnetic and **c** the hidden-order phases are schematically illustrated by the *dark ellipsoidal circles*

such domains is predominantly determined and strongly pinned by the underlying crystal conditions, such as internal stress or disorder.

3.4.3 Comparison with other Experimental and Theoretical Studies

Here we comment on the relevance of the present observation to other experimental and theoretical studies on the hidden-order phase of URu_2Si_2. As mentioned before, it might be difficult to directly detect the observed rotational symmetry breaking with other bulk experimental probes because of the domain structure.

3.4.3.1 Nuclear Magnetic Resonance (NMR)

Figure 3.25 shows the temperature dependence of the full-width at half-maximum (FWHM) for the ^{29}Si NMR resonance line measured at ambient pressure [19]. The NMR spectra exhibit an anomalous broadening in the hidden-order phase for $\mathbf{H} \parallel ab$, which has been reported by other groups as well [20, 22]. This phenomena has been discussed in terms of the orbital currents associated with the time-reversal symmetry breaking [23]. The present results may provide an alternative explanation to this NMR observation. The finite χ_{ab} below T_0 gives rise to the broadening of the NMR spectrum, which is estimated to be ~0.1 Oe at 4 T from the present results. Recently, Kambe et al. have found NMR linewidth oscillations as a function of applied magnetic

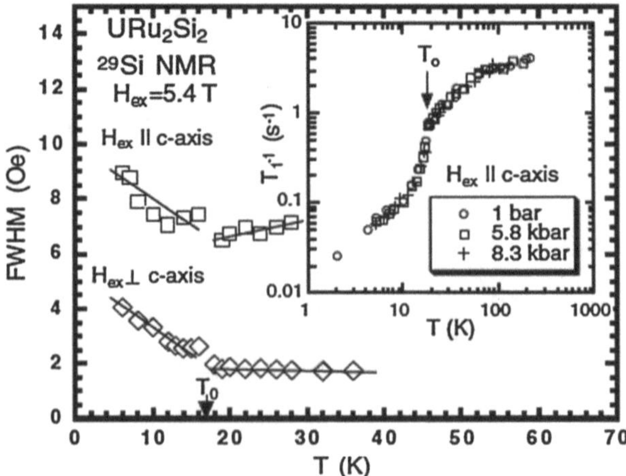

Fig. 3.25 Temperature dependence of the full-width at half-maximum (FWHM) for the ^{29}Si NMR resonance line measured at ambient pressure. A broadening is observed in the hidden-order phase. The *inset* shows the temperature dependence of T_1^{-1} measured at several pressures [19]

field orientation within the (001) plane in the hidden-order phase, which strongly supports the present 4-fold in-plane rotational symmetry breaking [24].

3.4.3.2 Nuclear Quadrupole Resonance (NQR)

We also note that some of the nuclear quadrupole frequency ν_Q measurements, which are sensitive to the local charge density, report a small change of $\nu_Q(T)$ at Ru site below T_0 [25]. About the crystal structure, until now no discernible lattice distortion has been reported below T_0. This may be due to weak coupling between the lattice and electronic excitations relevant to the 2-fold symmetry.

3.4.3.3 Cyclotron Resonance

Recently, Tonegawa et al. have performed a cyclotron resonance experiment, which directly measures the effective mass m^*_{CR} of electrons moving along external orbits on Fermi surface sheets through the simple relation $m^*_{\mathrm{CR}} = eH_{\mathrm{CR}}/\omega$, where H_{CR} is the resonance field and $\omega = 2\pi f$ is the microwave angular frequency [21]. They have conducted this technique on a high-quality $\mathrm{URu_2Si_2}$ single crystals (i.e. large $\omega_c\tau$), and find an anisotropic heavy band as well as several hole and electron bands that have been observed in previous quantum oscillation experiments. Figures 3.26a and b show the magnetic field dependence of the microwave dissipations measured at 28 and 60 GHz, respectively. In the 28-GHz measurement, a clear resonance peak contributed from heavy bands is signaled at high magnetic field of ~10 T. This newly found band can be assigned as the electron band centered at the M point (κ pocket). Furthermore, they find that the resonance peak from the main α hole band (labelled as D in Fig. 3.26a and b is split into two peaks by rotating field from [100] to [110] direction in the ab plane. The in-plane field angle dependence of the cyclotron mass is depicted in Fig. 3.26c. This behavior is unlikely to be attributed to the Fermi surface warping [21]. Rather, it is most naturally explained by the in-plane mass anisotropy which has heavy spots only near the orbit for $\mathbf{H} \parallel [110]$ as schematically shown in the inset of Fig. 3.26c. This breaks the 4-fold rotational symmetry in the ab-plane, consistent to our present results. In this case, there are two domains, one along [110] and the other along $[\bar{1}10]$ direction owing to the degenerate nature as observed in the present study. However, in contrast to the torque case in which signals of two domains cancel each other, the cyclotron resonance can detect both contributions even in a large single crystal.

3.4.3.4 Theories

The broken rotational symmetry imposes strong constraints on theoretical models which attempt to explain the hidden order. Recently, several theoretical models based on our observation of broken 4-fold symmetry have been proposed. Thalmeier and

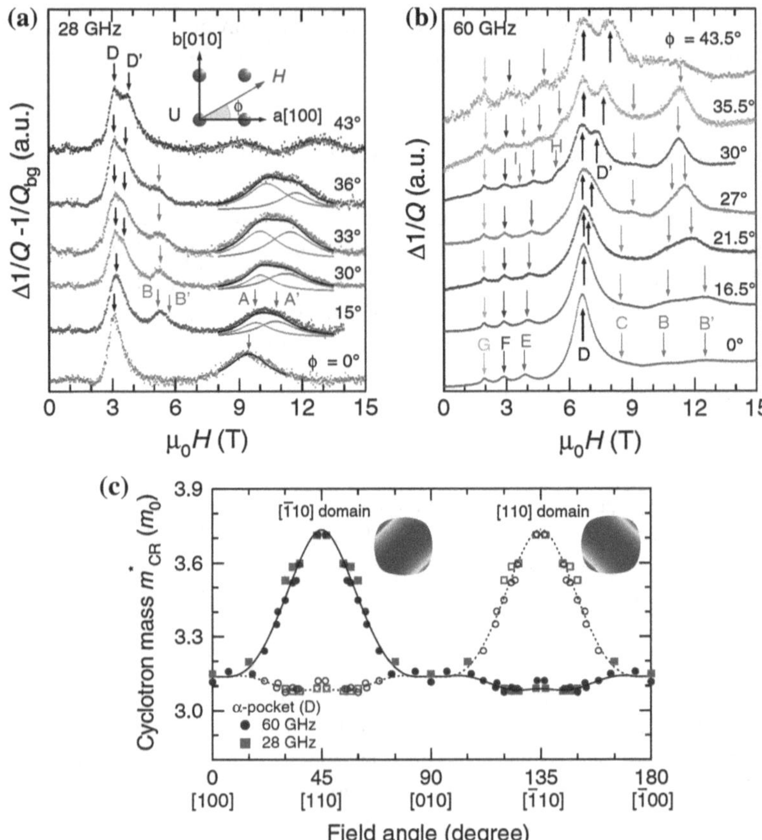

Fig. 3.26 Microwave dissipations as a function of magnetic field measured at **a** 28 and **b** 60 GHz. **c** Cyclotron effective mass of the α band as a function of the in-plane field angle expected for each domain [*solid* (open) symbols for the [$\bar{1}$10] ([110]) domain]. The *insets* show the schematic in-plane mass distribution of the α band [21]

Takimoto have performed the Landau theory analysis for the symmetry classification of the order parameters allowed in URu_2Si_2 [26]. The tetragonal symmetry group D_{4h} of URu_2Si_2 has three nontrivial irreducible representations (A_2, B_1 and B_2) and one doubly degenerate E representation, each of which has the even (+) and odd (−) parities with respect to time reversal. Such classification study has been studied by Santini and Amoretti [27], and several multipole order such as quadrupolar B_1 ($O_{x^2-y^2}$) [28] or B_2 (O_{xy}) [27–29]. In addition, higher-rank multipole order such as octapolar B_2 (T_{xyz}) [30] or hexadecapolar A_2 ($H_{xy(x^2-y^2)}$) [31] has been proposed so far.

Here the Landau free energy functional for such non-degenerate representations is expressed as [26]

$$B_1 : F = F_0 + g\eta(M_x^2 - M_y^2) + \tilde{g}\eta^2(M_x^2 + M_y^2), \tag{3.17}$$

$$B_2 : F = F_0 + g\eta M_x M_y + \tilde{g}\eta^2(M_x^2 + M_y^2), \tag{3.18}$$

$$A_2 : F = F_0 + g\eta M_x M_y(M_x^2 - M_y^2) + \tilde{g}\eta^2(M_x^2 + M_y^2), \tag{3.19}$$

where

$$F_0 = a\eta^2 + \frac{1}{2}b\eta^4. \tag{3.20}$$

Here η is an order parameter, M_i $(i = x, y)$ is the magnetization along i direction, and g, $\tilde{g}$ are constant. Note that the second terms in the right hand sides of Eqs. (3.17)–(3.19), which are linear in η, should be zero (i.e. $g = 0$) for antiferro-type ordering because of the translational invariance.

For the degenerate E representation, the Landau free energy is described as

$$F = F_0 + g_0(\eta_a^2 + \eta_b^2)(M_x^2 + M_y^2) \\ + g_1(\eta_a^2 - \eta_b^2)(M_x^2 - M_y^2) + 2g_2\eta_a\eta_b M_x M_y, \tag{3.21}$$

where

$$F_0 = a(\eta_a^2 + \eta_b^2) + \frac{1}{2}b_1(\eta_a^4 + \eta_b^4) + b_2\eta_a^2\eta_b^2. \tag{3.22}$$

Here $\eta = (\eta_a, \eta_b)$ is the 2-fold degenerate order parameter and g_0, g_1, g_2 are constant. In Eqs. (3.20) and (3.22), we have $a = a_0(T - T_0)$ and $a_0, b, b_1, b_2 > 0$. Furthermore, in magnetic fields, the magnetic part of the free energy F_m given by

$$F_m = \frac{\mathbf{M}^2}{2\chi_0} + \frac{1}{2}\lambda_1(M_x^4 + M_y^4) + \lambda_2 M_x^2 M_y^2 - \mathbf{M}\cdot\mathbf{H}, \tag{3.23}$$

where $\mathbf{M} = (M_x, M_y)$ and $\mathbf{H} = (H_x, H_y)$ are the magnetization and the magnetic field, λ_1, λ_2 are constant, and χ_0 is the background magnetic susceptibility measured for $\mathbf{H} \parallel ab$ planes, should be added, and then total free energy has to be minimized simultaneously with respect to the hidden order parameter and the induced magnetization.

The magnetic torque, which is obtained from the angle derivative of the free energy, for non-degenerate representations is given by

$$B_1 : \tau = -2\tau_0\Delta_0\left(1 - \frac{T}{\tilde{T}_0}\right)^{1/2}\sin 2\phi + \frac{1}{4}\tau_0\delta_0\chi_0 H^2 \sin 4\phi, \tag{3.24}$$

$$B_2 : \tau = \tau_0\Delta_0\left(1 - \frac{T}{\tilde{T}_0}\right)^{1/2}\cos 2\phi + \frac{1}{4}\tau_0\delta_0\chi_0 H^2 \sin 4\phi, \tag{3.25}$$

$$A_2 : \tau = \frac{1}{4}\tau_0\chi_0 H^2 \times \left[\delta_0 \sin 4\phi + 4\chi_0 \Delta_0 \left(1 - \frac{T}{\tilde{T}_0}\right)^{1/2} \cos 4\phi\right], \tag{3.26}$$

where

$$\tau_0 = \mu_0\chi_0 V H^2, \tag{3.27}$$

$$\delta_0 = 2\chi_0^2(\lambda_2 - \lambda_1), \tag{3.28}$$

$$\Delta_0 = g\chi_0\left(\frac{a_0\tilde{T}_0}{b}\right), \tag{3.29}$$

$$\tilde{T}_0 = T_0 - \frac{\tilde{g}\chi_0}{a_0}\chi_0 H^2. \tag{3.30}$$

Indeed these torque has 2- and 4-fold oscillational form. However, in quadrupolar B_1 and B_2 cases, the 2-fold oscillation becomes to be zero when the antiferro-type ordering appears ($g = 0$). This is a natural consequence because the neighboring quadrupoles with opposite signs induce no 4-fold symmetry breaking in such non-degenerate quadrupolar ordering. A ferro-type ordering (i.e. $g \neq 0$) may produce a 2-fold oscillation although antiferro-type one seems to be more appropriate in this material. In such a case, B_2-type quadrupolar ordering shows $\cos 2\phi$ oscillation, which agrees with our present results. However, the in-plane anisotropy in B_2-type quadrupolar ordering shows $(1 - T/\tilde{T}_0)^{1/2}$ dependence, incompatible with our results. In Fig. 3.27, we depict the observed 2-fold oscillation component $|A_{2\phi}|$ normalized by the sample volume V as a function of $1 - T/T_0$. Here the magnetic field dependence of T_0 is negligibly small [32], then $\tilde{T}_0 \simeq T_0$. The inset shows the temperature dependence of $|A_{2\phi}| \propto \chi_{ab}$ near T_0, which is well fitted with $\chi_{ab} \propto c(T_0 - T) + c'(T_0 - T)^2$. Therefore, the amplitude increase of $(1 - T/\tilde{T}_0)^{1/2}$ expected in the B_2-type quadrupolar ordering is too strong in comparison with our experiments. We also note that the hexadecapolar A_2-type ordering does not induce any 2-fold oscillation.

In degenerate representations, the magnetic torque is given by

$$\tau = 2\tau_0\Delta_0^{(0,1)}\left(1 - \frac{T}{\tilde{T}_0}\right)\sin 2\phi + \frac{1}{4}\tau_0\tilde{\delta}_0^{(0,1)}\chi_0 H^2 \sin 4\phi \tag{3.31}$$

for $E(0, 1)(0, zx)$ quadrupolar ordering and

$$\tau = 2\tau_0\Delta_0^{(1,1)}\left(1 - \frac{T}{\tilde{T}_0}\right)\cos 2\phi + \frac{1}{4}\tau_0\tilde{\delta}_0^{(1,1)}\chi_0 H^2 \sin 4\phi \tag{3.32}$$

for $E(1, 1)(yz, zx)$ quadrupolar ordering, where

$$\Delta_0^{(0,1)} = g_1\chi_0 \frac{a_0\tilde{T}_0}{b_1},\ \tilde{\delta}_0^{(0,1)} = 2\chi_0^2\left[(\lambda_2-\lambda_1)+\frac{2g_0^2}{b_1}\right], \tag{3.33}$$

$$\Delta_0^{(1,1)} = g_2\chi_0 \frac{a_0\tilde{T}_0}{b_1+b_2},\ \tilde{\delta}_0^{(1,1)} = 2\chi_0^2\left[(\lambda_2-\lambda_1)+\frac{2g_1^2}{b_1-b_2}\right]. \tag{3.34}$$

These models produce the 2-fold oscillation even in the case of antiferro-type ordering. In particular, the $E(1,1)(yz,zx)$ quadrupolar ordering shows a $\cos 2\phi$ oscillation, which has been observed in our experiments. Furthermore, the obtained temperature dependence of the in-plane anisotropy is $(1-T/\tilde{T}_0)$, also similar to our observation. The inset of Fig. 3.27 indicates that the observed in-plane anisotropy is well fitted with a function of $c(T_0-T)+c'(T_0-T)^2$ rather than only linear term. Such higher-order contribution will be given by including a higher-order term such as $(\eta_a^4-\eta_b^4)(M_x^2-M_y^2)$ in the free energy. This free energy analysis therefore concludes that the basic feature of our experiments is reproduced by the $E(1,1)$ quadrupolar ordering, which is also supported from analysis of resistive anisotropy [33]. Here, in this phase, there are four domains with $E(n_x,n_y)=E(\pm1,\pm1)$, and the sign of the 2-fold oscillation depends on the sign of $n_x n_y$. In large-volume samples, such 2-fold oscillation with opposite signs are cancelled, leading to small 2-fold oscillation amplitude. The symmetry of the hidden order parameter is then restricted to E^+-or E^--type, depending on whether the time reversal symmetry is preserved or not.

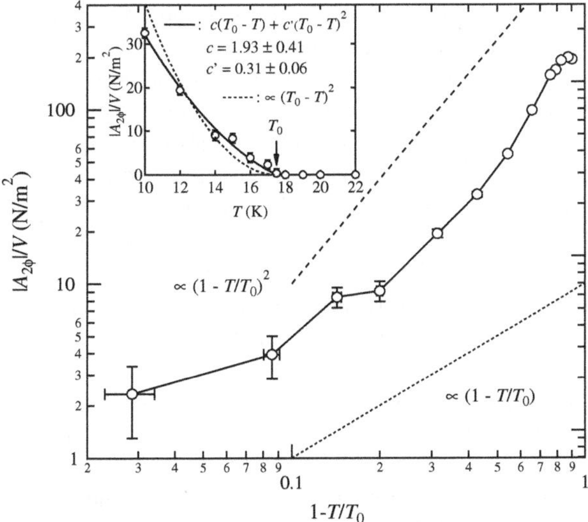

Fig. 3.27 The 2-fold oscillation amplitude $|A_{2\phi}|/V$ for sample #8 as a function of $1-T/T_0$ on a log–log scale. The *dotted* and *dashed lines* represent linear and quadratic temperature dependence, respectively. *Inset*: Temperature dependence of $|A_{2\phi}|/V$ for sample #8 near $T_0 = 17.5(1)$ K is fitted to $c(T_0-T)+c'(T_0-T)^2$ (*solid line*) and the quadratic dependence $\propto (T_0-T)^2$ (*dashed line*)

Recently, a spin nematic state is proposed as a candidate for the hidden order [34]. This ordering breaks the 4-fold symmetry in the spin space but time-reversal symmetry is preserved. This is categorized in the E^+ symmetry in the above representation. The temperature dependence of the in-plane anisotropy is explained by this scenario in a semi-quantitative way. Very recently, Ikeda et al. have performed ab initio calculations to evaluate the multipole correlations in URu_2Si_2 [35]. Based on the complete set of calculations for all the allowed symmetries, they conclude that the $E^-(D_{x(y)})$-type rank-5 dotriakontapolar order with broken time-reversal symmetry is the most stable at low temperatures in URu_2Si_2. Furthermore, such $E^-(D_{x(y)})$-type dotriakontapolar state is energetically close to $A_2^-(J_z)$-type antiferromagnetic one, leading to a natural explanation of the pressure-induced phase transition from the hidden-order to the antiferromagnetic phase. In their itinerant scenario, the Fermi surface nesting plays an essential role in the multipole fluctuations. They find that there is a nesting vector $\mathbf{Q}_c = (0, 0, 1)$ in the calculated paramagnetic Fermi surface of URu_2Si_2, and most importantly, the Fermi surface regions connected by this $\mathbf{Q}_c$ are mostly governed by $j_z = \pm 5/2$ components. In this case, one considers a subspace consisting of two components $j_z = \pm 5/2$, which can be mapped to pseudospin $\uparrow$ and $\downarrow$. Then the dipole J_z can be described by the (pseudospin) Pauli matrix σ_z, which has only diagonal components corresponding to $\pm 5/2 \Leftrightarrow \pm 5/2$. The dotriakontapole $D_{x(y)}$ is expressed as $\sigma_{x(y)}$, which possesses the off-diagonal elements to describe $\pm 5/2 \Leftrightarrow \mp 5/2$ transition. Note that this transition accompanies the angular momentum change of $5\hbar$, which is only allowed in rank 5. Now the staggered J_z state corresponds to the antiferromagnetic ordering along the c axis. On the other hand, the $D_{x(y)}$ state corresponds to the in-plane ordering, which indeed breaks 4-fold rotational symmetry in the *ab* plane. Therefore, the pressure-induced first-order phase transition from hidden-order to antiferromagnetic phase is regarded as a spin-flop transition of pseudospins from in-plane to out-of-plane direction. The calculated Fermi surface with the E^- state indeed breaks the in-plane 4-fold rotational symmetry [35], which may induce the experimentally observed in-plane magnetic anisotropy. The same ordered state has also been proposed by Rau and Kee [36], who found that the calculated temperature dependence of the in-plane anisotropy in this ordered state is consistent with our results. Another E^--type order parameter with broken 4-fold rotational and time-reversal symmetry has been proposed by Chandra et al. [37]. The experimental verifications on the time reversal symmetry breaking are thus highly desired.

We should note that an electronic state that breaks 4-fold rotational symmetry has also been suggested in some of the strongly electron correlated systems, including the pseudogap phase of high-T_c cuprates, and discussed in terms of stripe or nematic order [38, 39]. Recently such broken 4-fold rotational symmetry is also found in the Fe-based superconductors [40]. Quest for such an exotic order in the hidden-order phase deserves future studies.

3.4.3.5 Superconducting State

The symmetry of superconductivity, which is embedded in the hidden-order phase, should also be restricted by the found rotational asymmetry. Recently, the chiral d-wave superconducting state with the form of $k_x k_z + i k_y k_z$ has been proposed [41, 42]. The rotational symmetry breaking implies that the states $k_x k_z$ and $k_y k_z$ have different superconducting transition temperatures T_{c1} and T_{c2} ($<T_{c1}$), which results in a phase transition inside the superconducting phase at T_{c2}. Recently reported anomaly in the lower critical field measurements appears to support such exotic superconducting states [43], which will be discussed in Chap. 4

3.5 Summary

The transition into the hidden-order phase in the heavy-fermion superconductor URu_2Si_2 is one of the most intriguing phenomena among the strongly correlated electron systems. In present study, we have constructed the magnetic torque measurement system and investigated the out-of-plane and in-plane magnetic anisotropy in small pure single crystals very precisely. From this study, we find that the 4-fold rotational symmetry is broken below the hidden-order transition temperature. We summarize the results of our torque studies as follows.

- The out-of-plane torque measured with rotating field in the *ac* plane exhibits clear 2-fold oscillation and the temperature and magnetic field variations of the amplitudes are well described within the paramagnetic response. The out-of-plane torque measurements have also revealed that several crystals contain ferromagnetic impurities. We therefore use small pure single crystals which show no hysteresis behaviors.
- The in-plane torque measured with rotating field in the tetragonal *ab* plane shows 4-fold oscillation behavior and no 2-fold oscillation in the paramagnetic phase above $T_0 = 17.5$ K. Below T_0, on the other hand, the in-plane torque exhibits notable 2-fold oscillation which follows $\cos 2\phi$ form, indicating that an orthorhombicity with $\chi_{ab} \neq 0$ appears in the hidden-order phase.
- Extrinsic origins for the observed 2-fold oscillations such as a sample shape effect are excluded. Our torque study then concludes that the hidden-order phase is an electronic or magnetic state that breaks the tetragonal 4-fold rotational symmetry.

Our experimental observation captures a new essential feature of the hidden order parameter in URu_2Si_2, i.e. the low-temperature ordered phase is an electronic "nematic" state that breaks 4-fold rotational symmetry of the tetragonal crystal structure. Our study imposes a strong constraint on theoretical models that explains the nature of hidden-order phase and offer a new insight into various intriguing ordered phases in strongly correlated electron systems.

References

1. M. Herak, M. Miljak, A. Akrap, L. Forró, H. Berger, J. Phys. Soc. Jpn. **77**, 093701 (2008)
2. T. Sasaki, T. Fukase, Phys. Rev. B **59**, 13872 (1999)
3. C. Lupien, B. Ellman, P. Grütter, L. Taillefer, Appl. Phys. Lett. **74**, 451 (1999)
4. E. Ohmichi, T. Osada, Rev. Sci. Instrum. **73**, 3022 (2002)
5. H. Shishido, A.F. Bangura, A.I. Coldea, S. Tonegawa, K. Hashimoto, S. Kasahara, P.M.C. Rourke, H. Ikeda, T. Terashima, R. Settai, Y. Onuki, D. Vignolles, C. Proust, B. Vignolle, A. McCollam, Y. Matsuda, T. Shibauchi, A. Carrington, Phys. Rev. Lett. **104**, 057008 (2010)
6. D. Hall, E.C. Palm, T.P. Murphy, S.W. Tozer, Z. Fisk, U. Alver, R.G. Goodrich, J.L. Sarrao, P.G. Pagliuso, T. Ebihara, Phys. Rev. B **64**, 212508 (2001)
7. T.D. Matsuda, D. Aoki, S. Ikeda, E. Yamamoto, Y. Haga, H. Ohkuni, R. Settai, and Y. Onuki, J. Phys. Soc. Jpn. Suppl. A **77**, 362 (2008)
8. M.B. Maple, Y. Dalichaouch, B.W. Lee, C.L. Seaman, P.K. Tsai, P.E. Armstrong, Z. Fisk, C. Rossel, M.S. Torikachvili, Physica B **171**, 219 (1991)
9. A.P. Ramirez, T. Siegrist, T.T.M. Palstra, J.D. Garrett, E. Bruck, A.A. Menovsky, J.A. Mydosh, Phys. Rev. B **44**, 5392 (1991)
10. K. Hasselbach, P. Lejay, J. Flouquet, Phys. Lett. A **156**, 313 (1991)
11. C. Rossel, P. Bauer, D. Zech, J. Hofer, M. Willemin, H. Keller, J. Appl. Phys. **79**, 8166 (1996)
12. H. Amitsuka, K. Matsuda, I. Kawasaki, K. Tenya, M. Yokoyama, C. Sekine, N. Tateiwa, T.C. Kobayashi, S. Kawarazaki, H. Yoshizawa, J. Magn. Magn. Mater. **310**, 214 (2007)
13. S. Uemura, G. Motoyama, Y. Oda, T. Nishioka, N.K. Sato, J. Phys. Soc. Jpn. **74**, 2667 (2005)
14. V.G. Kogan, Phys. Rev. B **24**, 1572 (1981)
15. V.G. Kogan, M.M. Fang, S. Mitra, Phys. Rev. B **38**, 11958 (1988)
16. M. Angst, R. Puzniak, A. Wisniewski, J. Jun, S.M. Kazakov, J. Karpinski, J. Roos, H. Keller, Phys. Rev. Lett. **88**, 167004 (2002)
17. A. del Moral, J.I. Arnaudas, J.S. Cockaday, E.W. Lee, J. Magn. Magn. Mater. **40**, 331 (1984)
18. C. Broholm, J.K. Kjems, W.J. Buyers, P. Matthews, T.T. Palstra, A.A. Menovsky, J.A. Mydosh, Phys. Rev. Lett. **58**, 1467 (1987)
19. K. Matsuda, Y. Kohori, T. Kohara, K. Kuwahara, H. Amitsuka, Phys. Rev. Lett. **87**, 087203 (2001)
20. S. Takagi, S. Ishihara, S. Saitoh, H. Sasaki, H. Tanida, M. Yokoyama, H. Amitsuka, J. Phys. Soc. Jpn. **76**, 033708 (2007)
21. S. Tonegawa, K. Hashimoto, K. Ikada, Y.-H. Lin, H. Shishido, Y. Haga, T.D. Matsuda, E. Yamamoto, Y. Onuki, H. Ikeda, Y. Matsuda, T. Shibauchi, Phys. Rev. Lett. **109**, 036401 (2012)
22. O.O. Bernal, C. Rodrigues, A. Martinez, H.G. Lukefahr, D.E. MacLaughlin, A.A. Menovsky, J.A. Mydosh, Phys. Rev. Lett. **87**, 196402 (2001)
23. P. Chandra, P. Coleman, J.A. Mydosh, V. Tripathi, Nature **417**, 831 (2002) (London)
24. S. Kambe, Y. Tokunaga, H. Sakai, T.D. Matsuda, Y. Haga, Z. Fisk, R.E. Walstedt, Phys. Rev. Lett. **110**, 246406 (2013)
25. O.O. Bernal, M.E. Moroz, K. Ishida, H. Murakawa, A.P. Reyes, P.L. Kuhns, D.E. MacLaughlin, J.A. Mydosh, T.J. Gortenmulder, Physica B **378–380**, 574 (2006)
26. P. Thalmeier, T. Takimoto, Phys. Rev. B **83**, 165110 (2011)
27. P. Santini, G. Amoretti, Phys. Rev. Lett. **73**, 1027 (1994)
28. F.J. Ohkawa, H. Shimizu, J. Phys. Condens. Mater. **11**, L519 (1999)
29. H. Harima, K. Miyake, J. Flouquet, J. Phys. Soc. Jpn. **79**, 033705 (2010)
30. A. Kiss, P. Fazekas, Phys. Rev. B **71**, 054415 (2005)
31. K. Haule, G. Kotliar, Nature Phys. **5**, 796 (2009)
32. S.A.M. Mentink, T.E. Mason, S. Süllow, G.J. Nieuwenhuys, A.A. Menovsky, J.A. Mydosh, J.A.A.J. Perenboom, Phys. Rev. B **53**, R6014 (1996)
33. K. Miyake, J. Flouquet, J. Phys. Soc. Jpn. **79**, 035001 (2010)
34. S. Fujimoto, Phys. Rev. Lett. **106**, 196407 (2011)
35. H. Ikeda, M.-T. Suzuki, R. Arita, T. Takimoto, T. Shibauchi, Y. Matsuda, Nature Phys. **8**, 528 (2012)

36. J.G. Rau, H.-Y. Kee, Phys. Rev. B **85**, 245112 (2012)
37. P. Chandra, P. Coleman, R. Flint, Nature **493**, 621 (2013)
38. M. Vojta, Adv. Phys. **58**, 699 (2009)
39. R. Daou, J. Chang, D. LeBoeuf, O. Cyr-Choinière, F. Laliberté, N. Doiron-Leyraud, B.J. Ramshaw, R. Liang, D.A. Bonn, W.N. Hardy, L. Taillefer, Nature **463**, 519 (2010)
40. S. Kasahara, H.J. Shi, K. Hashimoto, S. Tonegawa, Y. Mizukami, T. Shibauchi, K. Sugimoto, T. Fukuda, T. Terashima, A.H. Nevidomskyy, Y. Matsuda, Nature **486**, 382 (2012)
41. Y. Kasahara, T. Iwasawa, H. Shishido, T. Shibauchi, K. Behnia, Y. Haga, T.D. Matsuda, Y. Onuki, M. Sigrist, Y. Matsuda, Phys. Rev. Lett. **99**, 116402 (2007)
42. K. Yano, T. Sakakibara, T. Tayama, M. Yokoyama, H. Amitsuka, Y. Homma, P. Miranovic, M. Ichioka, Y. Tsutsumi, K. Machida, Phys. Rev. Lett. **100**, 017004 (2008)
43. R. Okazaki, M. Shimozawa, H. Shishido, M. Konczykowski, Y. Haga, T.D. Matsuda, Y. Onuki, T. Shibauchi, Y. Matsuda, J. Phys. Soc. Jpn. **79**, 084705 (2010)

Chapter 4
Lower Critical Field Study on the Superconducting Phase

4.1 Introduction

In this chapter, we show the lower critical field study on the superconducting phase of URu_2Si_2. As described in the previous chapter, the hidden-order phase of this material is highly mysterious, highlighted by an absence of signature for long-range ordering in spite of distinct thermodynamic anomalies such as the specific heat and magnetic susceptibility. The superconductivity in this materials only appears in the hidden-order phase as seen in the pressure-temperature phase diagram, thus it is likely to be unusual.

Over the past 20 years, the study of such unconventional superconducting states has been become a central issue in condensed matter physics. Here, the full symmetry group $\mathscr{G}$ of the crystal consists of the gauge group $U(1)$, crystal point group G, spin rotation group $SU(2)$, and time reversal symmetry group $\mathscr{T}$,

$$\mathscr{G} = U(1) \otimes G \otimes SU(2) \otimes \mathscr{T}. \tag{4.1}$$

In conventional superconductors, only $U(1)$ gauge symmetry is broken below the transition temperature. Unconventional superconductivity, on the other hand, is defined as superconductors with additional symmetry breaking besides the $U(1)$ symmetry. In contrast to isotropic s-wave superconductivity, which is attributed to the electron-phonon interaction, such unconventional superconductors possess the anisotropic gap function with nodes along certain directions in the momentum space. Most importantly, such superconducting gap structure is intimately related to the pairing interaction, thus its determination is the most fundamental issue to clarify the underlying pairing mechanism.

R. Okazaki, *Hidden Order and Exotic Superconductivity in the Heavy-Fermion Compound URu_2Si_2*, Springer Theses, DOI: 10.1007/978-4-431-54592-7_4,

4.1.1 Anisotropic Superconducting Gap Structure

The superconducting gap function $\Delta^l_{\mathbf{s}_1,\mathbf{s}_2}(\mathbf{k})$ is proportional to the amplitude of the wave function for a Cooper pair $\psi^l_{\mathbf{s}_1,\mathbf{s}_2}(\mathbf{k}) = \langle c_{\mathbf{k},\mathbf{s}_1} c_{-\mathbf{k},\mathbf{s}_2}\rangle$, where $c_{\mathbf{k},\mathbf{s}_i}$ is the annihilation operator for electrons with momentum $\mathbf{k}$ and spin $\mathbf{s}_i$. In the simplest case with weak spin-orbit interaction, the total angular momentum $\mathbf{L}$ and total spin $\mathbf{S}$ are good quantum numbers, thus the gap function can be written as a product of the orbital and spin parts,

$$\Delta^l_{\mathbf{s}_1,\mathbf{s}_2}(\mathbf{k}) = g_l(\mathbf{k})\chi_s(\mathbf{s}_1, \mathbf{s}_2), \tag{4.2}$$

and the orbital part $g_l(\mathbf{k})$ is expanded using the spherical harmonics $Y_{lm}(\hat{\mathbf{k}})$ as

$$g_l(\mathbf{k}) = \sum_{m=-l}^{l} a_{lm}(k) Y_{lm}(\hat{\mathbf{k}}), \tag{4.3}$$

where $\hat{\mathbf{k}}$ is the unit vector that shows the direction of the Fermi wave vector $\mathbf{k}_F$. Here, the superconductor with $l = 0, 1, 2, \cdots$ are labelled as s, p, d-wave, $\cdots$, respectively.

The spin part of the gap function is expressed as a product of the spinors for two electrons in the Cooper pair and is described as a 2×2 matrix in the spin space,

$$\Delta^l_{\mathbf{s}_1,\mathbf{s}_2}(\mathbf{k}) = \begin{pmatrix} \Delta^l_{\uparrow\uparrow}(\mathbf{k}) & \Delta^l_{\uparrow\downarrow}(\mathbf{k}) \\ \Delta^l_{\downarrow\uparrow}(\mathbf{k}) & \Delta^l_{\downarrow\downarrow}(\mathbf{k}) \end{pmatrix} \tag{4.4}$$

In the spin singlet case, the spin part of the wavefunction is $|\uparrow\downarrow\rangle - |\downarrow\uparrow\rangle$, thus the gap function is given by

$$\Delta_s(\mathbf{k}) = i\Delta g_l(\mathbf{k})\sigma_y. \tag{4.5}$$

The quasiparticle energy dispersion is then given by

$$E_{\mathbf{k}} = \sqrt{\varepsilon^2_{\mathbf{k}} + \Delta^2 |g_l(\mathbf{k})|^2}, \tag{4.6}$$

where $\varepsilon_{\mathbf{k}}$ is the band energy relative to the chemical potential. For anisotropic pairing, therefore, the excitation energy depends on the form of $g_l(\mathbf{k})$, in contrast to the isotropic s-wave pairing case.

In the spin triplet case, the wave function possesses components corresponding to the three different spin projections on the quantization axis ($|\uparrow\uparrow\rangle, |\downarrow\downarrow\rangle, |\uparrow\downarrow\rangle + |\downarrow\uparrow\rangle$). Consequently, the order parameter can be written as

$$\Delta_t(\mathbf{k}) = i\mathbf{d}(\mathbf{k}) \cdot \boldsymbol{\sigma}\sigma_y = \begin{pmatrix} -d_x(\mathbf{k}) + id_y(\mathbf{k}) & d_z(\mathbf{k}) \\ d_z(\mathbf{k}) & d_x(\mathbf{k}) + id_y(\mathbf{k}) \end{pmatrix}, \tag{4.7}$$

and the dispersion is expressed as

$$E_{\mathbf{k}} = \sqrt{\varepsilon_{\mathbf{k}}^2 + |\mathbf{d}(\mathbf{k})|^2}. \tag{4.8}$$

These different energy dispersion relations provide us with vital information to determine the superconducting gap structure. For instance, in the fully-gapped s-wave and p-type axial and polar states, the energy dispersion is given by

$$E_{\mathbf{k}} = \begin{cases} \sqrt{\varepsilon^2 + \Delta_0^2} & \text{(Fully gapped)} \\ \sqrt{\varepsilon^2 + \Delta_0^2(k_x^2 + k_y^2)} & \text{(Axial)} \\ \sqrt{\varepsilon^2 + \Delta_0^2 k_z^2} & \text{(Polar)} \end{cases}, \tag{4.9}$$

where Δ_0 is a constant. In the fully gapped s-wave superconductor, the energy gap with a constant value opens over the Fermi surface. On the other hand, in the axial or polar state, there are nodes on the superconducting gap structure: In the axial state, the gap closes at $k_x = k_y = 0$, namely the point nodes locate on the poles of the Fermi surface. In the polar state, the gap is closed when $k_z = 0$, thus the node locates on the equator of the Fermi surface. These superconducting gap structures are schematically shown in Fig. 4.1. The density of states of quasiparticles for $E < \Delta_0$ is then given as

$$\frac{N(E)}{N_0} = \begin{cases} 0 & \text{(Fully gapped)} \\ \frac{1}{2}\frac{E}{\Delta_0} \log\left|\frac{E+\Delta_0}{E-\Delta_0}\right| & \text{(Axial)} \\ \frac{\pi}{2}\frac{E}{\Delta_0} & \text{(Polar)} \end{cases}, \tag{4.10}$$

where N_0 is the density of states in the normal state. The energy dependence of the density of states is shown in Fig. 4.2. At low energies, the density of states is proportional to E^2 and E for axial and polar states, respectively.

Experimentally, the superconducting gap structure has been clarified by probing these low-energy excitations using thermodynamic and transport measurements. In the fully-gapped s-wave superconductors, the energy gap has a finite constant value

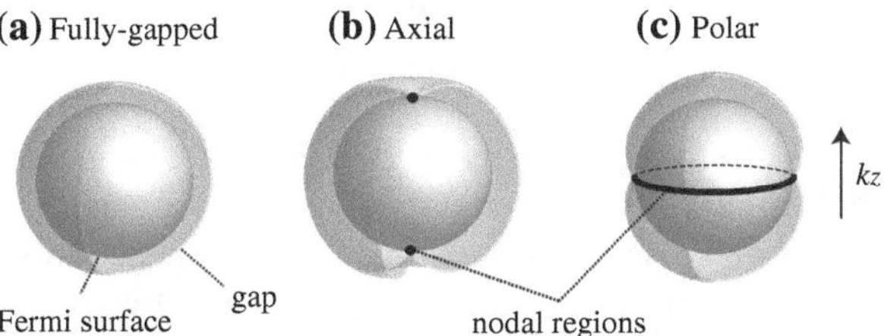

Fig. 4.1 Schematic figures for the superconducting gap structures for (**a**) fully-gapped, (**b**) axial, and (**c**) polar states

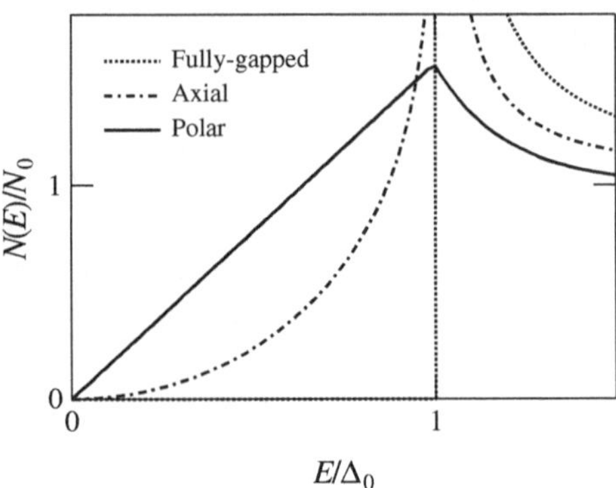

Fig. 4.2 Energy dependence of the density of states

over the Fermi surface, thus there is no quasiparticle state at $k_B T < \Delta_0$. Consequently, the thermodynamic and transport quantities exhibit an activation-type temperature dependence of $\exp(-\Delta_0/k_B T)$ at low temperatures. In contrast, unconventional superconductors have nodes in the gap function, which make the density of states finite at low energies. As discussed above, the density of states is proportional to $(E/\Delta_0)^n$ ($n = 1$ for the polar state and $n = 2$ for the axial state) at low energies, leading to the power law temperature dependence in the thermodynamic and transport quantities at $T \ll T_c$.

4.1.2 Penetration Depth and Superfluid Density

In the Meissner state, magnetic field is expelled from the superconductors. London introduced a phenomenological model of superconductivity in which the magnetic field inside a superconductor $\mathbf{B}$ obeys the equation,

$$\mathbf{B} - \lambda^2 \nabla^2 \mathbf{B} = 0, \tag{4.11}$$

where λ is the penetration depth given as

$$\lambda = \sqrt{\frac{mc^2}{4\pi n_s e^2}}. \tag{4.12}$$

Here m is an effective mass and n_s is a superfluid density, temperature dependence of which is strongly affected by the nodal topologies in the gap function.

Here we will describe a semiclassical model for the penetration depth and the superfluid density [1]. In a limiting case for $\lambda \gg \xi$, where ξ is the coherence length, the supercurrent $\mathbf{j}_s(\mathbf{r})$ is related to the vector potential $\mathbf{A}(\mathbf{r})$ as

$$\mathbf{j}_s(\mathbf{r}) = R\mathbf{A}(\mathbf{r}). \tag{4.13}$$

Here R is a response tensor given by

$$R_{ij} = \frac{e^2}{4\pi^3\hbar c}\oint_{\mathrm{FS}} dS \left[\frac{v_F^i v_F^j}{|\mathbf{v}_F|}\left(1 + 2\int_{\Delta}^{\infty}\frac{\partial f(E)}{\partial E}\frac{N(E)}{N_0}dE\right)\right], \tag{4.14}$$

where $f(E)$ is the Fermi function, $E = \sqrt{\varepsilon^2 + \Delta^2}$ is the quasiparticle energy and v_F^i are components of the Fermi velocity, $\mathbf{v}_F$. From Eqs. (4.11) and (4.13), one obtain

$$\mathbf{j}_s(\mathbf{r}) = \frac{c}{4\pi\lambda_{ii}^2}\mathbf{A}(\mathbf{r}). \tag{4.15}$$

Here we use a coordinate system defined by the principle axes of R, and the penetration depth is then given as

$$\lambda_{ii}^2 = \frac{c}{4\pi R_{ii}}, \tag{4.16}$$

and the superfluid density is described as

$$n_{ii} = \frac{cm_{ii}}{e^2}R_{ii}. \tag{4.17}$$

Then the normalized superfluid density is given as

$$\rho_{ii}(T) = \frac{n_{ii}(T)}{n_0} = \frac{R_{ii}(T)}{R_{ii}(0)} = \left[\frac{\lambda_{ii}(0)}{\lambda_{ii}(T)}\right]^2. \tag{4.18}$$

From Eqs. (4.14) and (4.18), the normalized superfluid density for a three-dimensional Fermi surface is given as,

$$\begin{pmatrix}\rho_{aa}(T)\\ \rho_{bb}(T)\end{pmatrix} = 1 - \frac{3}{4\pi T}\int_0^1 (1-z^2)\int_0^{2\pi}\begin{pmatrix}\cos^2\phi\\ \sin^2\phi\end{pmatrix} \times \int_0^{\infty}\cosh^{-2}\left(\frac{\sqrt{\varepsilon^2 + \Delta^2(T,\theta,\phi)}}{2T}\right)d\varepsilon d\phi dz, \tag{4.19}$$

and

$$\rho_{cc}(T) = 1 - \frac{3}{2\pi T}\int_0^1 z^2\int_0^{2\pi}\cos^2\phi \times \int_0^{\infty}\cosh^{-2}\left(\frac{\sqrt{\varepsilon^2 + \Delta^2(T,\theta,\phi)}}{2T}\right)d\varepsilon d\phi dz, \tag{4.20}$$

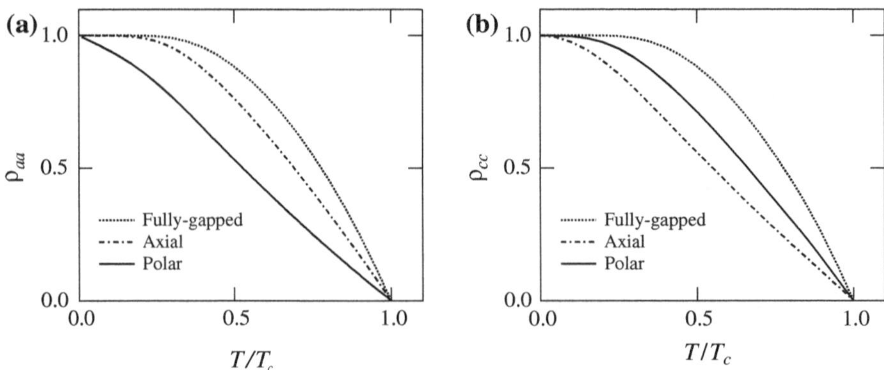

Fig. 4.3 Temperature dependence of the normalized superfluid density [(**a**) ρ_{aa} and (**b**) ρ_{cc}] for several gap structures with $\Delta_0 = 2k_B T_c$

where $z = \cos\theta$. The calculated superfluid densities are shown in Fig. 4.3. The superfluid density strongly depends on the nodal topologies as well as the gap amplitude. There is also significant difference in the anisotropy. Thus, the superfluid density provides vital information to determine the superconducting gap structure.

4.1.3 Lower Critical Field

Generally, superconductors are categorized as type-I or type-II by the magnitude of the Ginzbrug–Landau (GL) parameter $\kappa \equiv \lambda/\xi$. In type-I superconductors, weak magnetic field does not penetrate into the sample, that is the Meissner effect, but when it reaches the thermodynamic critical field H_c, the sample is no longer superconducting state. In type-II superconductors, below the lower critical field H_{c1}, the perfect diamagnetism is kept, similar to the case of type-I superconductors below H_c. In the mixed state ($H_{c1} < H < H_{c2}$), where H_{c2} is the upper critical field, because of negative surface energy of a wall separating normal and superconducting regions, a regular array of the vortices with the flux quantum ϕ_0 ($= ch/2e$) is formed to create as many domains as possible, resulting in spatially inhomogeneous superconductivity with the normal region in the cores of the vortex and the superconducting region in the surroundings of the cores.

Here, the magnetic field at which vortices first penetrate into superconductors is obtained from consideration of the Gibbs free energy function G,

$$G = n_L \zeta + \sum_{ij} U_{ij} - \frac{BH}{4\pi}, \tag{4.21}$$

where $n_L = B/\phi_0$ is the flux density and ζ is the line energy for one vortex, which is given as

$$\zeta = \left(\frac{\phi_0}{4\pi\lambda}\right)^2 \left(\ln\frac{\lambda}{\xi} + \varepsilon\right), \tag{4.22}$$

where ε is a numerical constant of the order of 0.1. The second term in Eq. (4.21) represents an interaction between vortices, which can be neglected at low magnetic fields, then

$$G \simeq B\left(\frac{\zeta}{\phi_0} - \frac{H}{4\pi}\right). \tag{4.23}$$

When $H < 4\pi\zeta/\phi_0$, G is an increasing function of B, then the lowest energy is obtained for $B = 0$, i.e., the Meissner state. On the other hand, when $H > 4\pi\zeta/\phi_0$, one can lower the energy by choosing finite B, then the vortex penetration occurs. Thus, the first penetration field (lower critical field) is given as

$$H_{c1} = \frac{4\pi\zeta}{\phi_0} = \frac{\phi_0}{4\pi\lambda^2}\left(\ln\frac{\lambda}{\xi} + \varepsilon\right). \tag{4.24}$$

In anisotropic materials, λ and ξ are given as λ_{aa} and ξ_{aa} for $\mathbf{H} \parallel c$ and as $\sqrt{\lambda_{aa}\lambda_{cc}}$ and $\sqrt{\xi_{aa}\xi_{cc}}$ for $\mathbf{H} \parallel a$. In Eq. (4.24), the logarithmic term $\ln(\lambda/\xi)$ is almost temperature-independent due to similar temperature dependence of λ and ξ. Thus, H_{c1} is proportional to λ^{-2}, which in turn is proportional to the superfluid density.

4.2 H_{c1} Measurement on $PrFeAsO_{1-y}$ Superconductor

As described in previous section, H_{c1} is directly related to the superfluid density, which provides us with an important information to determine the gap function. However, reliable determination of H_{c1} is a difficult task in the presence of the flux pinning and magnetic relaxation. Figure 4.4 schematically shows the internal flux density profile in a superconductor above H_{c1}. Vortices penetrate from the edge

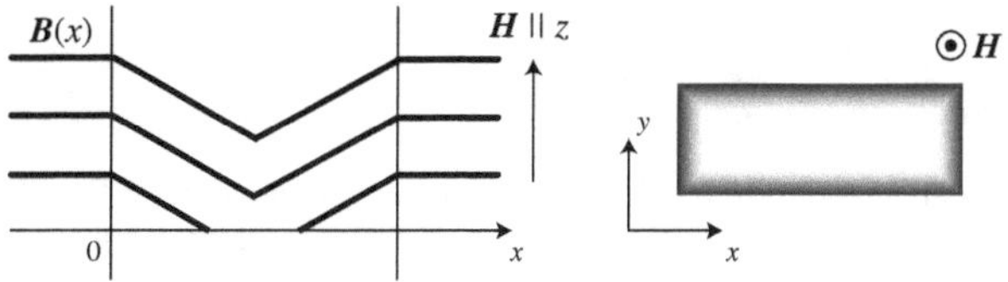

Fig. 4.4 Internal flux density profile in a type-II superconductor above H_{c1}. The *gray scale* in the *right panel* shows the magnetic induction B

of the crystal due to the pinning effects and then the magnetic profile becomes to be spatially inhomogeneous. Here, the standard technique to determine the lower critical field H_{c1} is to measure the dc magnetization in the bulk crystals. In this case, however, one obtains the averaged value over the crystal, from which the detection of the first flux penetration is very difficult.

In this section, we will describe how to precisely measure the lower critical field and recent our study on the Fe-based superconductor $PrFeAsO_{1-y}$. Figure 4.5a shows the temperature variation of the differential magneto-optics (MO) images of $PrFeAsO_{1-y}$ superconductor ($T_c \approx 34$ K) measured with small modulation field $\delta B = 0.2\,\mathrm{mT}$ in zero field [2]. The Meissner screening occurs completely within narrow temperature range. The flux penetrating behavior in a magnetic field measured at $T = 7.1\,\mathrm{K}$ is shown in Fig. 4.5b, which displays the magnetic field dependence of the MO images. The magnetic field distribution is well described by the Bean critical state model due to the flux pinning as schematically shown in Fig. 4.4, and this field distribution prevents the precise detection of the first flux penetration that occurs above H_{c1}.

To avoid these difficulties, we determined H_{c1} by directly detecting the positional dependence of the field H_p at which flux penetration occurs from the edge of the crystal [2]. Here we explain such measurement technique, which has been employed for H_{c1} determination in $PrFeAsO_{1-y}$ superconductor. The local magnetization near

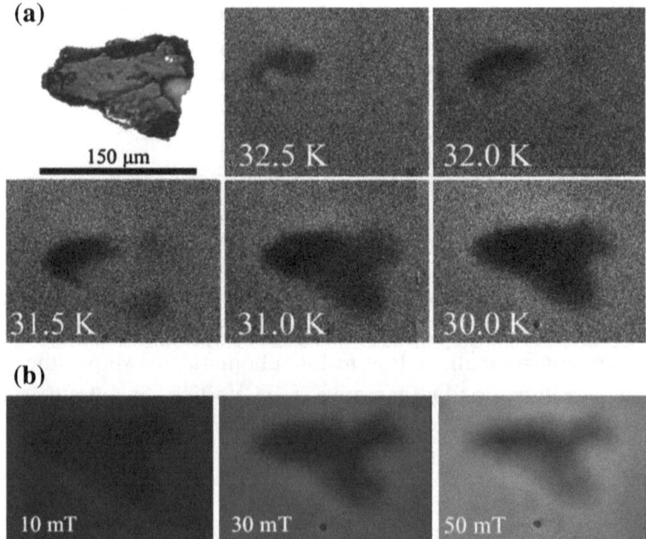

Fig. 4.5 **a** Differential magneto-optics (MO) images of superconducting $PrFeAsO_{1-y}$ ($T_c \approx 34\,\mathrm{K}$) with field modulation $\delta B = 0.2\,\mathrm{mT}$ in zero field. The Meissner screening occurs completely within narrow temperature range. The *left top panel* shows the picture of $PrFeAsO_{1-y}$ single crystal. **b** MO images measured at $T = 7.1\,\mathrm{K}$ in magnetic fields. Magnetic flux penetrates from the edge of the crystal and the field distribution shows the Bean critical state [2]

the surface of the platelet crystal has been measured by placing the sample on top of a miniature Hall-sensor array tailored in a GaAs/AlGaAs heterostructure, which was provided by Dr. Marcin Konczykowski at Ecole Polytechnique. Figure 4.6 displays the picture of miniature Hall-sensor array. There are 10 micro Hall sensors with an active area of $\approx 5 \times 5\,\mu m^2$ and the center-to-center distance of neighboring sensors is 20 μm (Fig. 4.6a). We apply the excitation current I_{ex} for long axis direction and measure the Hall voltage V_H of each sensor, giving the local magnetic induction $B_{\rm local} = R_H^{-1} V_H / I_{ex}$ at each position, where R_H is the Hall coefficient of the GaAs/AlGaAs Hall sensor.

Figure 4.7 shows typical curves of the square-root of local magnetic induction $\sqrt{B} \equiv \sqrt{\mu_0(M + \alpha H_a)}$ measured at the edge (circles) and at the center (squares) of the superconducting $PrFeAsO_{1-y}$ crystal with using the miniature Hall-sensor array [inset (a) of Fig. 4.7], plotted as a function of the external magnetic field H_a; the external field orientation $\mathbf{H} \parallel c$ and $T = 22\,\mathrm{K}$. The αH_a-term is obtained by a least squares fit of the low-field magnetization. The first penetration field $\mu_0 H_p$ corresponds to the field $\mu_0 H_p(\text{edge}) \simeq 2\,\mathrm{mT}$, above which $\sqrt{B}$ increases almost linearly, is clearly resolved. In Fig. 4.7, we show the equivalent curve, measured at the center of the crystal. At the center, $\sqrt{B}$ also increases linearly, but starts from a larger field, $\mu_0 H_p(\text{center}) \simeq 5\,\mathrm{mT}$.

We have measured the positional dependence of H_p and observed that it increases with increasing distance from the edge. To examine whether H_p(edge), i.e., H_p measured at $\leq$10 μm from the edge, truly corresponds to the field of first flux penetration at the boundary of the crystal, we have determined the local screening current density $j_{\rm edge} = \mu_0^{-1}(B_{\rm edge} - B_{\rm outside})/\Delta x$ at the crystal boundary. Here $B_{\rm edge}$ is the local magnetic induction measured by the sensor just inside the edge, $B_{\rm outside}$ is the induction measured by the neighboring sensor just outside the edge as shown in the inset of Fig. 4.7, and $\Delta x = 20\,\mu m$ is the center-to-center distance of neighboring sensors. For fields less than the first penetration field, $j_{\rm edge} \simeq \beta H_a$ is the Meissner current, which is simply proportional to the applied field (β is a constant determined

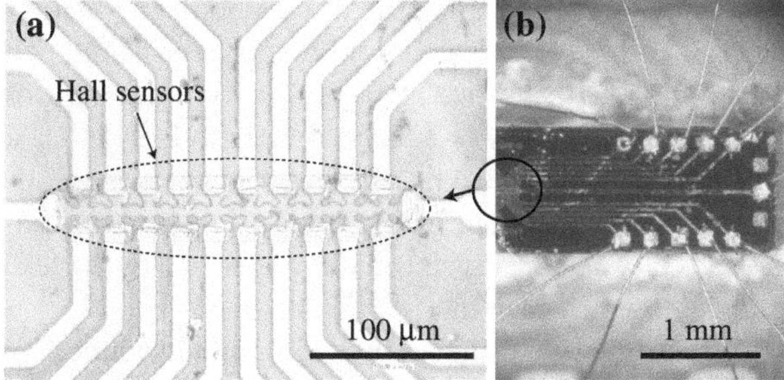

Fig. 4.6 **a** The miniature Hall-sensor array tailored in a GaAs/AlGaAs heterostructure. Each Hall sensor has an active area of $\approx 5 \times 5\,\mu m^2$; the center-to-center distance of neighboring sensors is 20 μm. **b** The Hall-sensor chip. The crystal is directly mounted on this chip using Apiezon N grease

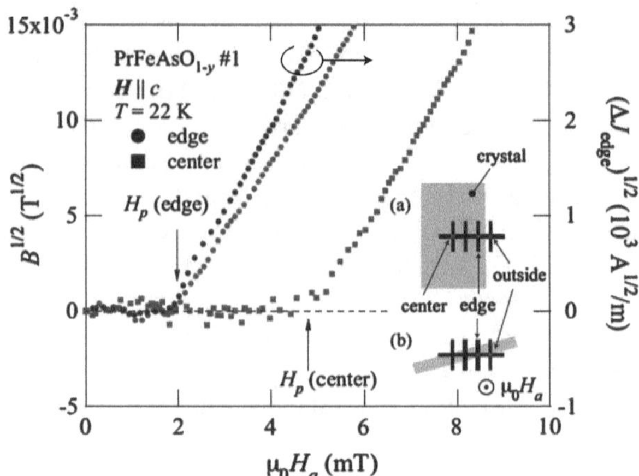

Fig. 4.7 Typical curves of the square-root of local magnetic induction $\sqrt{B} \equiv \sqrt{\mu_0(M + \alpha H_a)}$ (*left axis*) at the edge (*circles*) and at the center (*squares*) of the crystal and the square-root of change of the local screening current density $\sqrt{\Delta j_{\text{edge}}}$ (*right axis*) plotted as a function of H_a for $\mathbf{H} \parallel c$ at $T = 22$ K, in which H_a is increased after zero field cooling. The *insets* are schematic illustrations of the experimental setup for (*a*) $\mathbf{H} \parallel c$ and (*b*) $\mathbf{H} \parallel ab$-plane [2]

by geometry). At H_p, the screening current starts to deviate from linearity. Figure 4.7 shows the deviation $\Delta j_{\text{edge}} \equiv j_{\text{edge}} - \beta H_a$ as a function of H_a. As depicted in Fig. 4.7, $\sqrt{\Delta j_{\text{edge}}}$ again increases linearly with H_a above H_p(edge). This indicates that the H_p(edge) is very close to the true field of first flux penetration.

In Fig. 4.8, we compare the temperature dependence of H_p(edge) and H_p(center). In the whole temperature range, H_p(center) well exceeds H_p(edge). Moreover, H_p(center) increases with decreasing temperature without any tendency toward saturation. In sharp contrast, H_p(edge) saturates at low temperatures. The inset of Fig. 4.8 shows the difference between H_p measured in the center and at the edge: $\Delta H_p = H_p(\text{center}) - H_p(\text{edge})$. ΔH_p increases steeply with decreasing temperature. Also plotted in the inset of Fig. 4.8 is the remanent magnetization M_{rem} (i.e., the $H_a = 0$ value of M_{edge} on the decreasing field branch) measured at near the crystal center. This is proportional to the critical current density j_c arising from flux pinning. The temperature dependence of ΔH_p is very similar to that of j_c, which indicates that H_p(center) is strongly influenced by pinning. Hence, these results demonstrate that the lower critical field value determined by local magnetization measurements carried out at positions close to the crystal center is affected by vortex pinning effects and might be seriously overestimated.

The absolute value of H_{c1} is evaluated by taking into account the demagnetizing effect. For a platelet sample, H_{c1} is given by

$$H_{c1} = \frac{H_p}{\tanh\sqrt{0.36b/a}}, \tag{4.25}$$

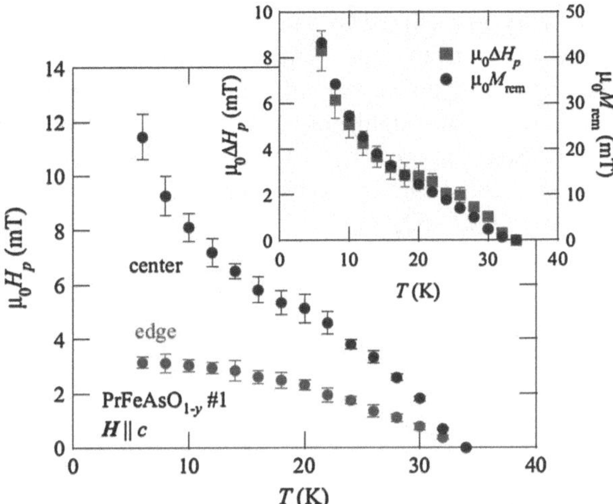

Fig. 4.8 Temperature dependence of the flux penetration fields H_p at the edge and the center of the crystal. The *inset* shows the temperature dependence of the difference between H_p in the center and at the edge (*left axis*), as well as the remanent magnetization M_{rem} (*right axis*) [2]

where a and b are the width and the thickness of the crystal, respectively [4]. In the situation where $\mathbf{H} \parallel c, a = 63\,\mu\text{m}$ and $b = 18\,\mu\text{m}$, while $a = 18\,\mu\text{m}$ and $b = 63\,\mu\text{m}$ for $\mathbf{H} \parallel ab$-plane. These values yield $H_{c1}^{c} = 3.22H_p$ and $H_{c1}^{ab} = 1.24H_p$, respectively. In Fig. 4.9, we plot H_{c1} as a function of temperature both for $\mathbf{H} \parallel c$ and $\mathbf{H} \parallel ab$-plane. The solid line in Fig. 4.9 indicates the temperature dependence

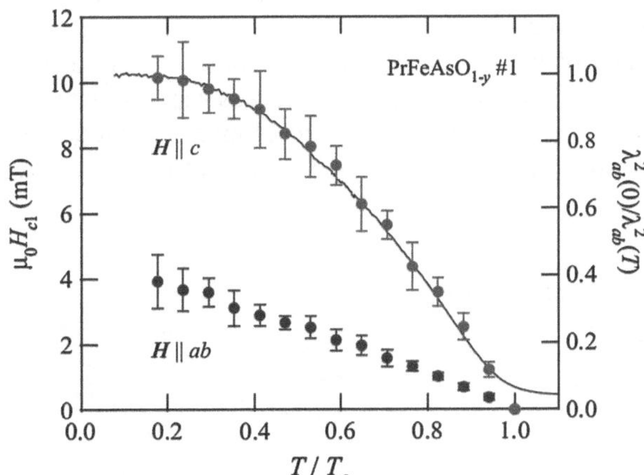

Fig. 4.9 Lower critical fields as a function of temperature in $PrFeAsO_{1-y}$ single crystals (*left axis*) [2]. The *solid line* (*right axis*) presents the superfluid density determined by surface impedance measurements on crystals from the same batch [3]

of the superfluid density normalized by the value at $T = 0\,\mathrm{K}$, which is obtained from ab-plane penetration depth measurements of a sample from the same batch [3]. $H_{c1}^{c}(T)$ is well scaled by the superfluid density, which is consistent with fully gapped superconductivity. To roughly estimate the in-plane penetration depth at low temperatures, we use the approximate single-band London formula [Eq. (4.24)], yielding $\lambda_{ab} \sim 280\,\mathrm{nm}$. This value is in close correspondence with the μSR results in slightly underdoped LaFeAs(O,F) [5].

4.3 Experiment

4.3.1 Samples

We used high-quality single crystals of URu_2Si_2 with a large residual resistivity ratio of $RRR = 670$, which were grown by the Czochralski pulling method in a tetra-arc furnace as described in previous chapter. The well defined superconducting transition was confirmed by the specific heat measurements. Experiments have been performed on a single crystal with the dimension of $2.3 \times 0.75 \times 0.15\,\mathrm{mm}^3$, the picture of which is shown in Fig. 4.10.

4.3.2 H_{c1} Measurement on URu_2Si_2

In this study, we first investigated the pinning properties in the crystals by measuring the magnetic field distribution in the vortex state, which is obtained by the scanning micro Hall sensor with an active area of $2 \times 2\,\mu\mathrm{m}^2$. The Hall sensor was scanned on the surface area of the crystal by using the piezoelectric device. The distance between the Hall sensor and the crystal surface was kept constant by monitoring the tunneling currents. Next we measured the local magnetic induction very precisely by using a miniature Hall-sensor array with each active area of $\approx 5 \times 5\,\mu\mathrm{m}^2$ (the

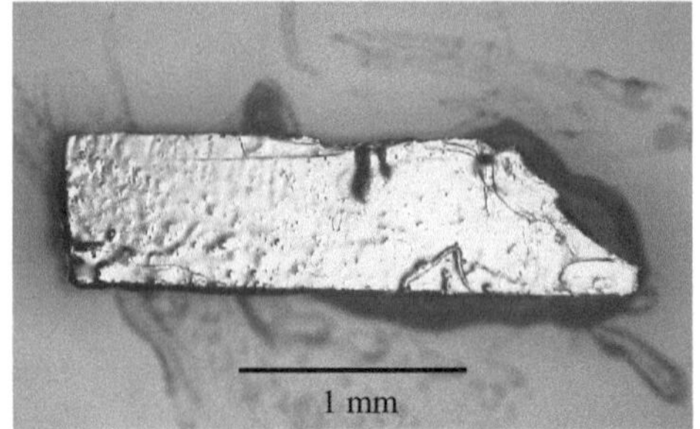

Fig. 4.10 URu_2Si_2 single crystal for the lower critical field measurement. The top shiny cleaved surface is the ab plane. The sample is surrounded by Apiezon N grease

center-to-center distance of neighboring sensors is 20 μm) as shown in Fig. 4.6. The crystal was directly placed on top of the array with a small amount of grease (Apiezon N grease). In both measurements, earth magnetic field was shielded by μ-metal. The residual field at the sample space is less than 0.5 μT. In all measurements, the external field was applied after the sample was zero-field cooled to the desired temperature from the temperature above T_c.

We explain the experimental setting as schematically shown in Fig. 4.11. All measurement systems are controlled by the personal computer via the LabVIEW program. The Hall resistance is measured by an ac resistance bridge LR-700 with low excitation current $I = 0.3\,\mu\text{A}$ in a ^{3}He cryostat with handmade copper coil and in a dilution refrigerator with superconducting magnet. The temperatures are monitored and controlled by CryoCon 62.

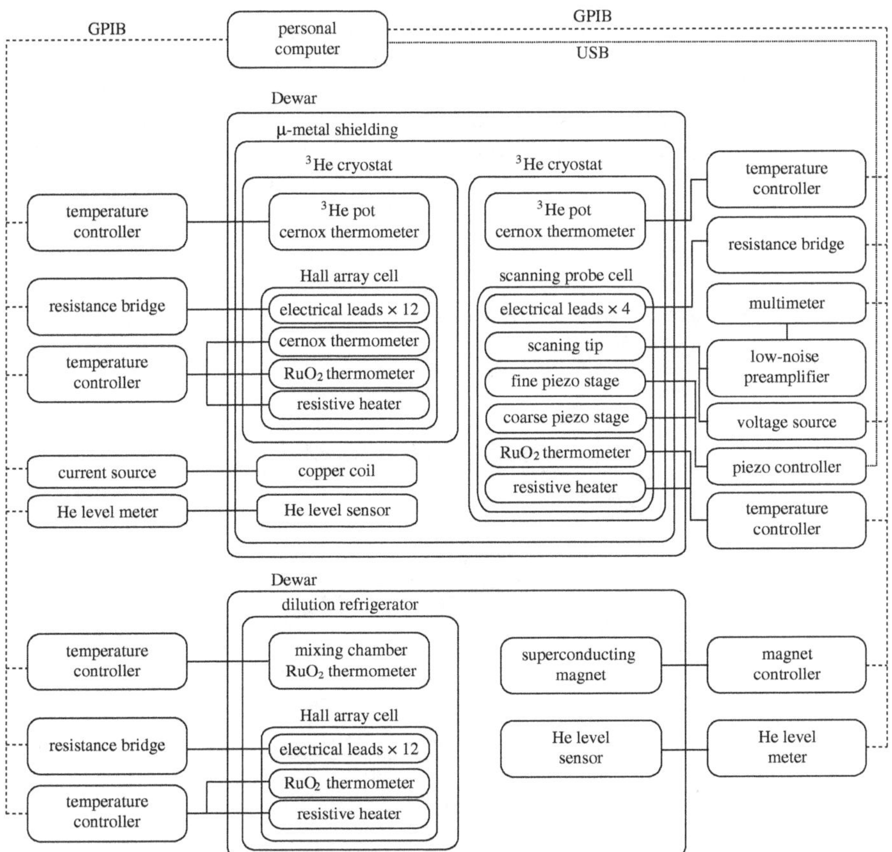

Fig. 4.11 Schematic view of the whole experimental system

4.4 Results

4.4.1 Magnetic Field Distribution in the Meissner and Vortex States

Figure 4.12a displays the line profile of the magnetic induction near the edge of *ab* plane in the Meissner state when the external magnetic field $\mu_0 H = 0.05\,\mathrm{mT}$ is applied parallel to the *c* axis. The profile is measured by the scanning Hall-probe microscopy. With approaching from the outside of the crystal, the magnetic induction is enhanced to $B \sim 0.08\,\mathrm{mT}$ followed by a sharp drop near the edge of the crystal. The magnetic induction is perfectly screened inside the crystal. Now we define the edge at the peak position of the local magnetic induction, as indicated by the dotted line. The enhancement of the magnetic induction at the edge originates from the magnetic field induced by the shielding currents flowing the edge region [6]. This demonstrates the homogeneous superconducting state of the crystal.

Figure 4.12b shows the spatial distribution of the local magnetic induction at several external magnetic fields associated with the flux penetration at $T = 0.5\,\mathrm{K}$. The magnetic induction shows a monotonic decay with the distance *d* from the edge. This indicates the Bean critical state dominated by the bulk flux pinning rather than

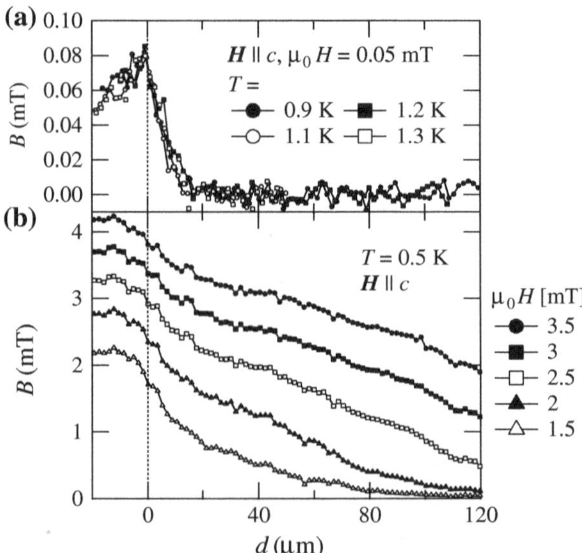

Fig. 4.12 The positional dependence of the local magnetic induction *B* near the edge region of the crystal, which is determined by the scanning Hall-probe microscopy. The *dashed line* is the position of the edge. **a** The profile of magnetic induction in the Meissner state. The external magnetic field ($\mu_0 H = 0.05\,\mathrm{mT}$) is applied parallel to the *c* axis. **b** The profile of magnetic induction in the vortex state at several fields at $T = 0.5\,\mathrm{K}$

by the Bean-Livingston surface barrier. In this case, H_{c1} is determined from the local magnetic induction near the edge, at which the first flux penetration is most sensitively detected [2]. This is in contrast to the case of surface pinning, in which the magnetic flux distribution shows a dome-like shape and the flux penetration field can be detected at the center of the crystal [7].

4.4.2 Local Magnetic Induction

We next measure the local magnetic induction near the edge B_{edge} by using the Hall-sensor array for $\mathbf{H} \parallel a$ and $\mathbf{H} \parallel c$ to determine the first penetration field H_p in the bulk pinning case. The sample edge is located between the Hall sensors 1 and 2 as shown in Fig. 4.13c, which is confirmed by the Meissner response of each sensor, as shown in Fig. 4.12a. In fact, sensor 2 exhibits a perfect Meissner response, while slight enhancement of B from $\mu_0 H$ is observed in sensor 1. Figure 4.13a, b displays the field dependence of B_{edge} measured by sensor 2 for $\mathbf{H} \parallel a$ and $\mathbf{H} \parallel c$, respectively. The flux penetration fields H_p shown by the triangles are clearly resolved by the deviation from the Meissner state ($B_{\text{edge}} = 0$).

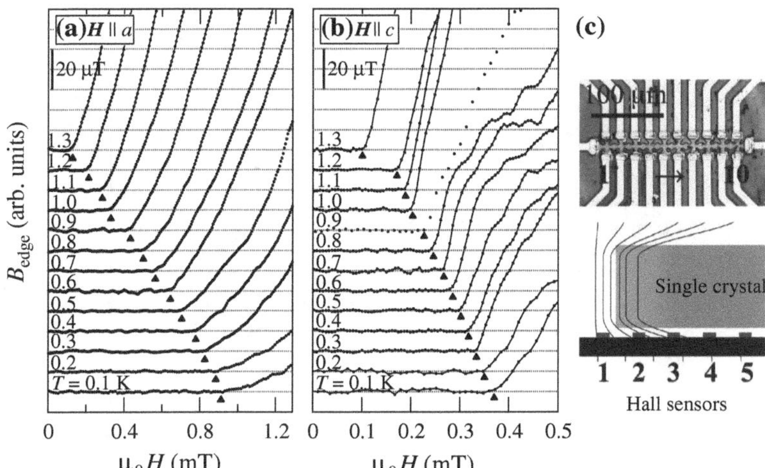

Fig. 4.13 **a** Local magnetic induction B_{edge} measured by the sensor at the edge of the crystal (Sensor 2), as a function of H for $\mathbf{H} \parallel a$. The *triangles* indicate the flux penetration field H_p. Each data is vertically shifted for clarity. Sensor 1 locates just outside of the crystal. **b** The same plot for $\mathbf{H} \parallel c$. **c** Photograph of the miniature Hall-sensor array, and a schematic illustration of the experimental setup

4.4.3 Lower Critical Field

Figure 4.14 depicts the temperature dependence of H_{c1}^{a} and H_{c1}^{c}, where H_{c1}^{a} and H_{c1}^{c} are the lower critical fields for $\mathbf{H} \parallel a$ and $\mathbf{H} \parallel c$, respectively. We evaluate $H_{c1}^{c} = 3.82 H_p$ and $H_{c1}^{a} = 1.15 H_p$ by using the expression [Eq. (4.25)] which takes into account the demagnetization effect [4]. The magnetic penetration depth is evaluated from H_{c1} to $\lambda_a \simeq 0.8\,\mu$m, which is quantitatively consistent with the μSR results of $\lambda_a = 0.7$–$1\,\mu$m [8, 9]. According to the Ginzburg-Landau theory, the temperature variation of H_{c1} anisotropy, $\gamma_{H_{c1}} \equiv H_{c1}^{c}/H_{c1}^{a}$, may be different from that of H_{c2} in multiband superconductors [10, 11]. At $T \to T_c$, however, $\gamma_{H_{c1}}$ should coincide with the H_{c2} anisotropy. The experimentally determined $\gamma_{H_{c1}} \simeq 3$ close to T_c is then fully consistent with the reported H_{c2} anisotropy near T_c [12], giving us confidence in the accuracy of the result. As seen in Fig. 4.14, H_{c1}^{a} increases linearly below $\sim$0.8 K with decreasing temperature. A cusp structure can be seen at $T_p = 1$ K. Meanwhile, H_{c1}^{c} increases steeply below T_c and increases linearly after exhibiting a distinct change in the slope with a kink at $T_Q = 1.2$ K. As shown by the dotted lines, H_{c1}^{c} at low temperatures increases linearly down to 55 mK, while H_{c1}^{a} exhibits a tendency to saturation below 200 mK.

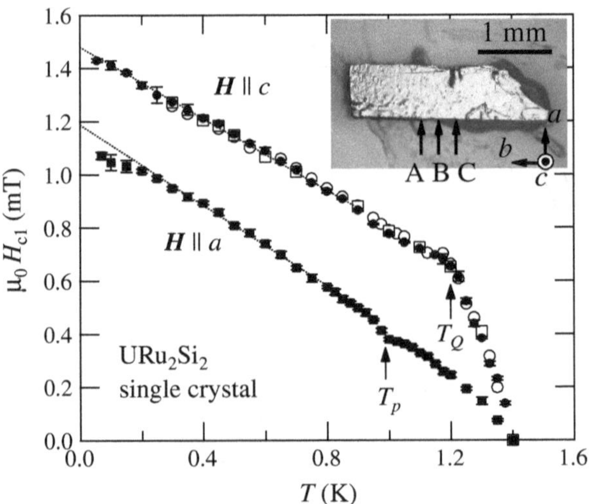

Fig. 4.14 *Inset* Photograph of the single crystal URu_2Si_2 used in the present study. Local magnetic induction is measured at the edge regions A, B, and C. *Main panel* Temperature dependence of the lower critical fields H_{c1} for $\boldsymbol{H} \parallel a$ (*solid squares*) and $\boldsymbol{H} \parallel c$ (*solid circles*), which are determined at the edge region B shown in the inset. *Arrows* indicate a kink anomaly at $T_Q = 1.2$ K and a cusp behavior at $T_p \simeq 1$ K. The *dotted lines* are the fits to the linear dependence. *Open circles* and *open squares* show $H_{c1}^{c}(T)$ determined at the edge regions A and C shown in the *inset*, respectively

4.5 Discussion

4.5.1 Temperature Dependence of H_{c1}

First we discuss the temperature variation of H_{c1} in the low temperature regime. Since H_{c1} is proportional to the superfluid density, the T-linear dependence of H_{c1}^c down to low temperatures indicates the presence of line nodes in the superconducting gap function. Moreover, the tendency to saturation in $H_{c1}^a(T)$ below 200 mK indicates that the line nodes are located parallel to the *ab* planes (horizontal node). This is because the supercurrents always flow parallel to the nodal planes for $\mathbf{H} \parallel c$, while for $\mathbf{H} \parallel a$ supercurrents have a component which flows across the horizontal nodes. This component is perpendicular to the velocities of the quasiparticles around the nodes, which reduces the contributions of nodal quasiparticles to the superfluid density. Since in the multiband superconductors the penetration depth and hence the lower critical field are governed by the band with light mass, it is natural to consider that the horizontal line node locates in the spherical light hole band. This is consistent with the previous results of angle-resolved thermal conductivity [13].

Most remarkable features found in the present study are anomalies at T_Q for $\mathbf{H} \parallel c$ and T_p for $\mathbf{H} \parallel a$. In the specific heat measurements, no anomaly has been observed at T_p in the present crystals with large *RRR* values. However, an extrinsic peak structure has been found in the vicinity of T_p for low-quality crystals with *RRR* < 100, possibly due to the inhomogeneous distribution of T_c [14]. Thus we cannot rule out a possibility that the observed T_p-anomaly in H_{c1}^a originates from a tiny portion with low-T_c phase in the crystal. On the other hand, no anomaly has been reported in the heat capacity at T_Q for any crystals, indicating no evidence of the low-T_c phase as an origin for the T_Q-anomaly. It is also unlikely that the anomaly at T_Q arises from an inhomogeneous flux penetration of the crystal, because of the following reasons. First, the open squares and circles in Fig. 4.14 show H_{c1}^c determined at different edge regions in the same crystal. The distinct kink anomaly at T_Q is quite well reproduced. Second, as shown in Fig. 4.12a, the Meissner shielding currents flow homogeneously and the crystal has well defined edge. These results indicate that the anomaly of $H_{c1}^c(T)$ at T_Q is an intrinsic property. Note that, in the crystals with lower *RRR* values, the anomaly at T_Q has never been reported [15].

4.5.2 Two-Gap Model

We here try to fit the temperature variations of H_{c1}^a and H_{c1}^c by using a multiband model. In recent angle-resolved thermal conductivity and specific heat measurements, two distinct superconducting gap structures having different nodal topology with horizontal line nodes in the hole band and point nodes in the electron have been resolved [13, 16, 17]. To examine whether multiband effect can explain anomalous behaviors of the observed lower critical fields, we discuss the temperature variations of H_{c1}^a and H_{c1}^c in terms of the two-gap model below. In the two-band model, the in-plane and out-of-plane superfluid density normalized by their values at $T = 0\,\text{K}$, n_s^a and n_s^c, respectively, can be expressed as

$$n_s^i(T) = x^i n_h^i(T) + (1 - x^i) n_e^i(T), \quad i = a, c \tag{4.26}$$

where n_h^i and n_e^i are the normalized superfluid density of hole and electron bands, respectively. x^i is the ratio of the electron and hole contributions given by

$$x^i = \frac{X_h^i}{X_h^i + X_e^i}, \quad X_{h(e)}^i = \int \frac{(v_{F,h(e)}^i)^2}{|\mathbf{v}_{F,h(e)}|} dS_{h(e)}, \tag{4.27}$$

where $\mathbf{v}_{F,h(e)}$ is the Fermi velocity of holes (electrons) [18]. The hole mass obtained by the dHvA measurements is nearly isotropic $m_h^a \approx m_h^c \approx 13m_0$ [19, 20], where m_h^a (m_h^c) is the mass of the hole along the a (c) axis and m_0 is the free electron mass. The heavy electron mass is anisotropic and is estimated to be $m_e^a \approx 85m_0$ and $m_e^c \approx 305m_0$ from the large Sommerfeld coefficient in the heat capacity ($\gamma \sim 80\,\text{mJ/K}^2\text{mol}$) [21] and the anisotropy of upper critical field H_{c2} [12]. Then we obtain $x^a \approx 0.87$ and $x^c \approx 0.95$. Theses values are close to unity, and therefore the superfluid density, particularly its temperature variation at low temperatures, should be governed by the light hole band. This again provides a strong support to the horizontal line nodes located on the hole bands as mentioned before. Furthermore, the presence of point nodes has been suggested along the c axis in the heavy electron bands [16, 17]. Then, we calculate the superfluid density by assuming the gap functions $\Delta_h(\mathbf{k}, T) = \Delta_h(T) \times 2\sin\theta\cos\theta$ with point and horizontal line nodes for hole bands and $\Delta_e(\mathbf{k}, T) = \Delta_e(T) \times \sin\theta$ with c-axis point nodes for electron bands, where θ is the polar angle measured from the c axis.

The normalized lower critical fields, $h^{a,c}(T) \equiv H_{c1}^{a,c}(T)/H_{c1}^{a,c}(0)$, are related to the superfluid density as $h^c = n_s^a$ and $h^a = \sqrt{n_s^a n_s^c}$. First we try to fit $h^a(T)$. Because the hole band dominates at low temperatures, $\Delta_h(\mathbf{k}, T)$ is unambiguously determined by $h^a(T)$. The best fit is obtained by $\Delta_h(0) = 1.6k_B T_c$ and $\Delta_e(0) = 4.0k_B T_c$. As shown by the thick line in Fig. 4.15a, the fitting result reproduces well the overall temperature dependence of $h^a(T)$, including the observed tendency toward saturation at low temperatures. We note that the $\Delta_h(0)$ value is close to $\Delta_h(0) \sim \hbar v_{F,h}/\pi\xi_h \sim 1.5k_B T_c$, which is obtained from $v_{F,h} \sim 2 \times 10^4\,\text{m/s}$ [19] and $\xi_h \sim 25\,\text{nm}$. Here $\xi_h = \sqrt{\Phi_0/2\pi\mu_0 H_{c2}^h}$ is the coherence length of the hole band estimated from the "virtual upper critical field" of the hole band $\mu_0 H_{c2}^h \sim 0.5\,\text{T}$ [16]. In Fig. 4.15b, $h^c(T)$ calculated by using the same $\Delta_h(0)$ and $\Delta_e(0)$ values is plotted by the thick line. In sharp contrast to $h^a(T)$, the calculation strongly deviates from the data. We have tried to fit the data by assuming various $\Delta_h(0)$ and $\Delta_e(0)$ values and other gap symmetries, however could not reproduce the data, particularly the anomaly at T_Q (Table 4.1).

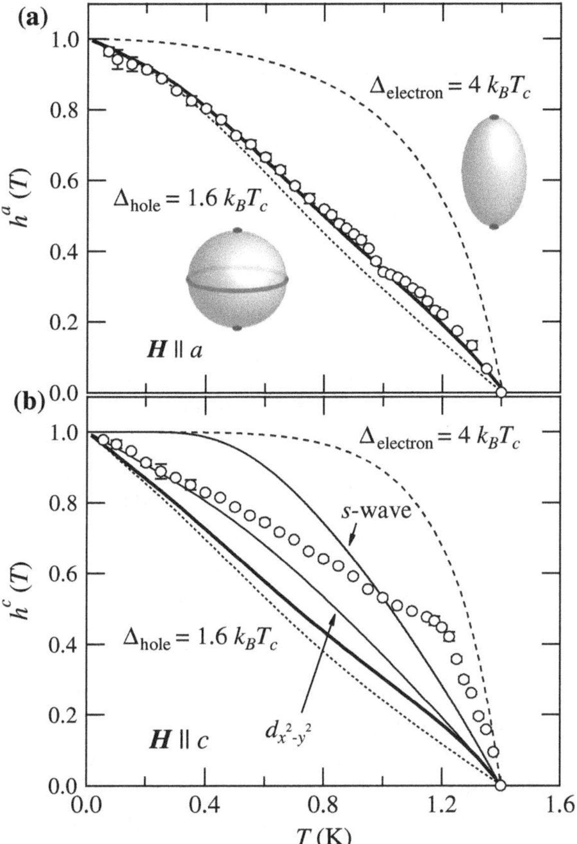

Fig. 4.15 Comparison of $h^{a,c} \equiv H_{c1}^{a,c}(T)/H_{c1}^{a,c}(0)$ (symbols) with (**a**) $\sqrt{n_s^a n_s^c}$ and (**b**) n_s^a, respectively. The *dotted* and *dashed lines*indicate the superfluid densities for the hole band with $\Delta_h = 1.6k_BT_c$ and the electron band with $\Delta_e = 4.0k_BT_c$, respectively, which are obtained by assuming the illustrated nodal topologies suggested by the thermal conductivity measurements [16]. The *solid thick* and *thin lines* represent the results of the two-gap fitting and the calculations for other gap symmetries, respectively

4.5.3 Anomaly at T_Q

On the basis of the present results, it is concluded that the multigap effect cannot be an origin of the anomaly at T_Q. Here we note that a change in the slope of $H_{c1}(T)$ has also been observed in the heavy-fermion superconductors UPt_3 [22] and $(U_{1-x}Th_x)Be_{13}$ [23, 24]. However, there are crucial differences from our observation. First, in URu_2Si_2 $H_{c1}(T)$ is strongly suppressed below T_Q, while in UPt_3 and $(U_{1-x}Th_x)Be_{13}$ $H_{c1}(T)$ is enhanced below the kink temperature. Second, in URu_2Si_2 the kink anomaly is observed solely for $\mathbf{H} \parallel c$, while in UPt_3 and $(U_{1-x}Th_x)Be_{13}$ the anomaly is observed in any field directions. In the latter compounds, the kink anomaly

Table 4.1 Fitting parameters for the two-band model

	Axis	m	Weight for n_s	$\Delta_i(0)$	$g(\mathbf{k})$
Hole band	a	$13m_0$	0.87	$1.6k_B T_c$	$\sin\theta\cos\theta$
	c	$13m_0$	0.95		
Electron band	a	$85m_0$	0.13	$4.0k_B T_c$	$\sin\theta$
	c	$305m_0$	0.05		

$g(\mathbf{k})$ denotes the angular part of the gap function

is attributed to the transition into another superconducting phase with different gap structure.

The observed strong suppression of $H_{c1}^c(T)$ implies that a vortex penetration to the sample is drastically changed at T_Q. Here we qualitatively explain this anomalous behavior of $H_{c1}(T)$ in terms of a peculiar vortex dynamics associated with chiral domains due to the multicomponent superconducting order parameter with broken time-reversal symmetry (TRS), which has been proposed by the thermal conductivity study recently [16]. Such order parameter consists of two components $k_z k_X$ and $ik_z k_Y$, where X and Y denote [110] and [1$\bar{1}$0] directions, respectively. In tetragonal URu_2Si_2, these two components are essentially same since $X = Y$. In this case, those pairing states must have the same transition temperature. However, recent magnetic torque measurements have revealed that the four-fold rotational symmetry is broken in the hidden-order phase [25], and it is natural to consider that the superconducting order parameter couples to the hidden order, because the superconductivity appears only in the hidden-order phase as shown by recent pressure studies [26]. These experimental results then imply that the transition temperature of the $k_z k_X$ state may differ from that of the $k_z k_Y$ state. In this case, the pairing state without broken TRS is stable just below T_c, and the other transition to the pairing state with broken TRS occurs at a lower critical temperature. Assuming that this second transition temperature is T_Q, the order parameter with broken TRS can produce domain walls along the c axis at T_Q. Then magnetic fields can penetrate inside through the domain wall even in the Meissner state as suggested in Ref. [27]. Thus the formation of domain walls induces a peculiar reduction of the penetration field only for $\mathbf{H} \parallel c$. As shown in Fig. 4.12, our scanning Hall probe detects no apparent inhomogeneous field distribution, which implies that such domains, if exist, are smaller than the size of Hall sensor ($\sim$2 μm). We also mention that the kink anomaly at T_Q has never been observed in the crystals with low *RRR* values [15], implying that the domain formation is sensitive to the impurities. It should be noted that a possible presence of such a domain structure, which influences the vortex dynamics, has been suggested in UPt_3, $U_{1-x}Th_xBe_{13}$, Sr_2RuO_4 and $PrOs_4Sb_{12}$ [28, 29]. We also note that the second superconducting transition discussed here is similar to the transition from the p-wave pairing state with TRS to the chiral p-wave state without TRS which has been proposed for Sr_2RuO_4 [30].

4.6 Summary

In this study, we have determined accurately the lower critical fields H_{c1} of ultraclean URu_2Si_2. In general, the determination of H_{c1} is very difficult due to the vortex pinning. Here we utilizes and develops a new technique to precisely measure H_{c1} by using local probes which enable us to evaluate the positional dependence of the local magnetic induction. By using this technique, we have firstly evaluated the temperature dependence of H_{c1} in Fe-based superconductor $PrFeAsO_{1-y}$, which is in agreement with that of the superfluid density determined from the penetration depth measurements. This experimental fact justifies our H_{c1} measurement method using the local probes.

We have utilized this technique to determine the lower critical fields of URu_2Si_2, and find that H_{c1} exhibits several distinct features, which have never been observed in any other superconductors. The temperature dependence of H_{c1} at low temperatures suggests the horizontal line nodes in the energy gap in the light hole band. We show that the whole $H_{c1}(T)$ for $\mathbf{H} \parallel a$ can be explained by the two gaps with line and point nodes, which is consistent with the previous reports. Furthermore the obtained energy gap values well agrees with other thermodynamic and transport measurements. In sharp contrast, for $\mathbf{H} \parallel c$ we find a distinct kink behavior in the slope of $H_{c1}(T)$ at 1.2 K, which cannot be accounted for by the two gaps. This anomalous low-field diamagnetic response has been discussed in the light of the multicomponent order parameter with broken time-reversal symmetry.

References

1. B.S. Chandrasekhar, D. Einzel, Ann. Phys., Lpz. **2**, 535 (1993)
2. R. Okazaki, M. Konczykowski, C.J. van der Beek, T. Kato, K. Hashimoto, M. Shimozawa, H. Shishido, M. Yamashita, M. Ishikado, H. Kito, A. Iyo, H. Eisaki, S. Shamoto, T. Shibauchi, Y. Matsuda, Phys. Rev. B **79**, 064520 (2009)
3. K. Hashimoto, T. Shibauchi, T. Kato, K. Ikada, R. Okazaki, H. Shishido, M. Ishikado, H. Kito, A. Iyo, H. Eisaki, S. Shamoto, Y. Matsuda, Phys. Rev. Lett. **102**, 017002 (2009)
4. E.H. Brandt, Phys. Rev. B **60**, 11939 (1999)
5. H. Luetkens, H.-H. Klauss, M. Kraken, F.J. Litterst, T. Dellmann, R. Klingeler, C. Hess, R. Khasanov, A. Amato, C. Baines, J. Hamann-Borrero, N. Leps, A. Kondrat, G. Behr, J. Werner, B. Buechner, Nat. Mater. **8**, 305 (2009)
6. E. Zeldov, John R. Clem, M. McElfresh, M. Darwin, Phys. Rev. B **49**, 9802 (1994)
7. T. Shibauchi, M. Konczykowski, C.J. van der Beek, R. Okazaki, Y. Matsuda, J. Yamaura, Y. Nagao, Z. Hiroi, Phys. Rev. Lett. **99**, 257001 (2007)
8. E.A. Knetsch, A.A. Menovsky, G.J. Nieuwenhuys, J.A. Mydosh, A. Amato, R. Feyerherm, F.N. Gygaxb, A. Schenck, R.H. Heffner, D.E. MacLaughlin, Physica B **186**, 300 (1993)
9. A. Amato, Rev. Mod. Phys. **69**, 1119 (1997)
10. V.G. Kogan, Phys. Rev. B **66**, 020509(R) (2002)
11. P. Miranovic, K. Machida, V.G. Kogan, J. Phys. Soc. Jpn. **72**, 221 (2003)
12. R. Okazaki, Y. Kasahara, H. Shishido, M. Konczykowski, K. Behnia, Y. Haga, T.D. Matsuda, Y. Onuki, T. Shibauchi, Y. Matsuda, Phys. Rev. Lett. **100**, 037004 (2008)
13. Y. Kasahara, H. Shishido, T. Shibauchi, Y. Haga, T.D. Matsuda, Y. Onuki, Y. Matsuda, New J. Phys. **11**, 055061 (2009)

14. T.D. Matsuda, D. Aoki, S. Ikeda, E. Yamamoto, Y. Haga, H. Ohkuni, R. Settai, Y. Onuki, J. Phys. Soc. Jpn. **77**(Suppl. A), 362 (2008)
15. S. Wüchner, N. Keller, J.L. Tholence, J. Flouquet, Solid State Commun. **85**, 355 (1993)
16. Y. Kasahara, T. Iwasawa, H. Shishido, T. Shibauchi, K. Behnia, Y. Haga, T.D. Matsuda, Y. Onuki, M. Sigrist, Y. Matsuda, Phys. Rev. Lett. **99**, 116402 (2007)
17. K. Yano, T. Sakakibara, T. Tayama, M. Yokoyama, H. Amitsuka, Y. Homma, P. Miranovic, M. Ichioka, Y. Tsutsumi, K. Machida, Phys. Rev. Lett. **100**, 017004 (2008)
18. R. Prozorov, R.W. Giannetta, Supercond. Sci. Technol. **19**, R41 (2006)
19. H. Ohkuni, Y. Inada, Y. Tokiwa, K. Sakurai, R. Settai, T. Honma, Y. Haga, E. Yamamoto, Y. Onuki, H. Yamagami, S. Takahashi, T. Yanagisawa, Philo. Mag. B **79**, 1045 (1999)
20. H. Shishido, K. Hashimoto, T. Shibauchi, T. Sasaki, H. Oizumi, N. Kobayashi, T. Takamasu, K. Takehana, Y. Imanaka, T.D. Matsuda, Y. Haga, Y. Onuki, Y. Matsuda, Phys. Rev. Lett. **102**, 156403 (2009)
21. M.B. Maple, J.W. Chen, Y. Dalichaouch, T. Kohara, C. Rossel, M.S. Torikachvili, M.W. McElfresh, J.D. Thompson, Phys. Rev. Lett. **56**, 185 (1986)
22. E. Vincent, J. Hammann, L. Taillefer, K. Behnia, N. Keller, J. Flouquet, J. Phys. Condens. Matter **3**, 3517 (1991)
23. R.H. Heffner, J.L. Smith, J.O. Willis, P. Birrer, C. Baines, F.N. Gygax, B. Hitti, E. Lippelt, H.R. Ott, A. Schenck, E.A. Knetsch, J.A. Mydosh, D.E. MacLaughlin, Phys. Rev. Lett. **65**, 2816 (1990)
24. U. Rauchschwalbe, F. Steglich, G.R. Stewart, A.L. Giorgi, P. Fulde, K. Maki, Europhys. Lett. **3**, 751 (1987)
25. R. Okazaki, T. Shibauchi, H.J. Shi, Y. Haga, T.D. Matsuda, E. Yamamoto, Y. Onuki, H. Ikeda, Y. Matsuda, Science **331**, 439 (2011)
26. H. Amitsuka, K. Matsuda, I. Kawasaki, K. Tenya, M. Yokoyama, C. Sekine, N. Tateiwa, T.C. Kobayashi, S. Kawarazaki, H. Yoshizawa, J. Magn. Magn. Mater. **310**, 214 (2007)
27. M. Ichioka, Y. Matsunaga, K. Machida, Phys. Rev. B **71**, 172510 (2005)
28. E. Dumont, A.C. Mota, Phys. Rev. B **65**, 144519 (2002)
29. T. Cichorek, A.C. Mota, F. Steglich, N.A. Frederick, W.M. Yuhasz, M.B. Maple, Phys. Rev. Lett. **94**, 107002 (2005)
30. D.F. Agterberg, Phys. Rev. Lett. **80**, 5184 (1998)

Chapter 5
Vortex Lattice Melting Transition

5.1 Introduction

In this chapter, we will show the thermal- and magneto-transport study in the vortex mixed state of URu_2Si_2. In general, magnetic flux lines take a periodic Abrikosov lattice arrangement in order to minimize the repulsive energy between themselves. However, in the compounds that have large thermal fluctuations and quenched disorder such as high-T_c cuprates, the vortex states could be varied to liquid or glass phase, where each vortex can move freely or be frozen at random, occupying a large part of H-T phase diagram. In high-T_c cuprates with the large anisotropy, the pancake and Josephson vortices having different shapes from an ordinary vortex which reflect the layered structures are realized. We stress that one of the most fascinating phenomena of those "vortex matter" is the first-order vortex lattice melting transition from the vortex lattice to the vortex liquid phase. In this case, the mean-field transition line $H_{c2}(T)$ marks only a crossover line and the thermodynamic phase transition line is shifted to the melting line $H_m(T)$. Through our transport study, we find such vortex melting transition phenomena in URu_2Si_2 even at subkelvin temperatures.

5.1.1 Thermal Fluctuations

We first discuss thermal fluctuations in superconductors, which are neglected in the mean-field approach but important for the vortex states in the high-T_c cuprates. We assume that the superconducting order parameter ψ indeed fluctuates by an amount $\delta\psi$ due to thermal effect. Since the fluctuations are expected to be random, $\langle\delta\psi\rangle = 0$, but $\langle(\delta\psi)^2\rangle > 0$. In the critical region $\langle(\delta\psi)^2\rangle \geq \langle|\psi|\rangle^2$, where we cannot treat the fluctuations as the perturbation, the Ginzbrug–Landau (GL) theory breaks down. Let us consider the Gaussian fluctuation, which neglects the $|\psi|^4$ term in the GL free energy density f,

$$f = f_n + \alpha(T)|\psi|^2 + \gamma|\nabla\psi|^2. \tag{5.1}$$

R. Okazaki, *Hidden Order and Exotic Superconductivity in the Heavy-Fermion Compound URu2Si2*, Springer Theses, DOI: 10.1007/978-4-431-54592-7_5,

Here f_n is the free energy density in the normal state and the magnetic field is taken to be zero. The Fourier transform Eq. (5.1) leads to

$$\Delta f_q = f_q - f_{nq} = \alpha(T)|\psi_q|^2 + \gamma q^2|\psi_q|^2 = \gamma\left(\frac{\alpha(T)}{\gamma} + q^2\right)|\psi_q|^2. \qquad (5.2)$$

We can use equipartition theorem for each q to perform a thermal average of Eq. (5.2). Thus,

$$\langle \Delta f_q \rangle = \left\langle \gamma\left(\frac{\alpha}{\gamma} + q^2\right)|\psi_q|^2\right\rangle = \gamma\left(\frac{\alpha}{\gamma} + q^2\right)\langle|\psi_q|^2\rangle \approx k_B T_c, \qquad (5.3)$$

where

$$\langle|\psi_q|^2\rangle = \frac{k_B T_c}{\gamma\left(q^2 + \alpha/\gamma\right)}. \qquad (5.4)$$

The critical region $\langle(\delta\psi)^2\rangle \geq \langle|\psi|\rangle^2$ can be estimated from the summation of the fluctuations, which is given by [1, 2]

$$\frac{|T - T_c|}{T_c} \leq \frac{1}{2}\left(\frac{k_B T_c}{H_c(0)^2\xi(0)^3}\right)^2 \equiv G_i, \qquad (5.5)$$

where $H_c(0)$ is the thermodynamic critical field at zero temperature. Here G_i denotes the Ginzburg number, that is the fundamental parameter governing the strength of thermal fluctuations in the above phenomena. It shows the relative size of the thermal energy $k_B T_c$ and the condensation energy within the coherence volume. For anisotropic superconductors, G_i is enhanced by the anisotropy parameter $\varepsilon \equiv (m_c/m_{ab})^{1/2} > 1$ [3],

$$G_i = \frac{1}{2}\left(\frac{\varepsilon k_B T_c}{H_c(0)^2\xi_a(0)^3}\right)^2. \qquad (5.6)$$

In conventional low-T_c superconductors, G_i ranges from 10^{-11} to 10^{-7}, while in high-T_c cuprates with long penetration depth, short coherence length, and high transition temperature, it is as large as 10^{-2} [3]. In zero field, the region given by Eq. (5.5) is extremely small even in high-T_c cuprates. However, in magnetic fields sufficiently strong, the superconducting fluctuations acquire an effective one-dimensional (1D) character along the field direction. This reduction of the effective dimensionality enhances the fluctuations.

5.1.2 Vortex Lattice Melting

Each vortex in the periodic lattice fluctuates due to thermal effects, leading to the vortex liquid state with increasing temperature. A number of authors suggested the existence of the melting transition between a vortex lattice and a vortex liquid [4–7], and by analogy with the melting transition in solids, one might expect that it would be the first-order thermodynamic phase transition, at which the long range order would be destroyed discontinuously [8]. However, in conventional low-T_c superconductors with less thermal fluctuations, it is very difficult to observe the melting transition experimentally because the melting line $H_m(T)$ takes place in the immediate vicinity of the mean-field transition line $H_{c2}(T)$ [9].

One simple model to describe the melting transition is given by the Lindemann criterion, in which the vortex lattice melts when a mean-square thermal vibration amplitude of each vortex $\langle \delta x^2 \rangle$ reaches $c_L^2 a_\triangle^2$. Here $c_L \approx 0.1$ and $a_\triangle = \left(\frac{2}{\sqrt{3}} \frac{\phi_0}{B}\right)^{1/2}$ are the Lindemann number and the intervortex distance of a triangular lattice, respectively. Minimizing the vibration energy, which is the sum of the elastic displacement energy and the energy required to stretch the flux line against the line tension, and equating it with $\sim k_B T$, one obtain the vibration amplitude given by [8]

$$\langle \delta x^2 \rangle = k_B T \left(\frac{4\pi^2 \lambda^2}{\phi_0^{3/2} B^{1/2}} \right). \tag{5.7}$$

Using the Lindemann criterion, the melting transition line (H_m, T_m) is obtained as

$$k_B T_m = \frac{c_L^2 \phi_0^{5/2}}{4\pi^2 [\lambda(T_m)]^2} B^{-1/2}, \tag{5.8}$$

$$H_m(T) \propto (T_c - T)^2. \tag{5.9}$$

Another estimation of the melting temperature is performed by Larkin and Varlamov. They determined T_m by the equality of the free energies of lattice and liquid phases as

$$T_m - T_c(H) = 2y \left(\frac{H}{\tilde{H}_{c2}(0)} \right)^{\frac{2}{3}} \sqrt[3]{G_i/\varepsilon^2}\, T_c(H), \tag{5.10}$$

where $y \approx -7$ for the 3D case and $y \approx -10$ for the 2D case [10]. Here $\tilde{H}_{c2}(0)$ is a linear extrapolation of the initial slope of $H_{c2}(T)$ at $T_c(H = 0)$ to $T \to 0\,\text{K}$.

5.1.3 Experimental Evidence of Melting Transition

The distinct vortex lattice melting in high-T_c cuprates with large G_i, which expands the vortex liquid region in the H-T phase diagram, has been observed in various

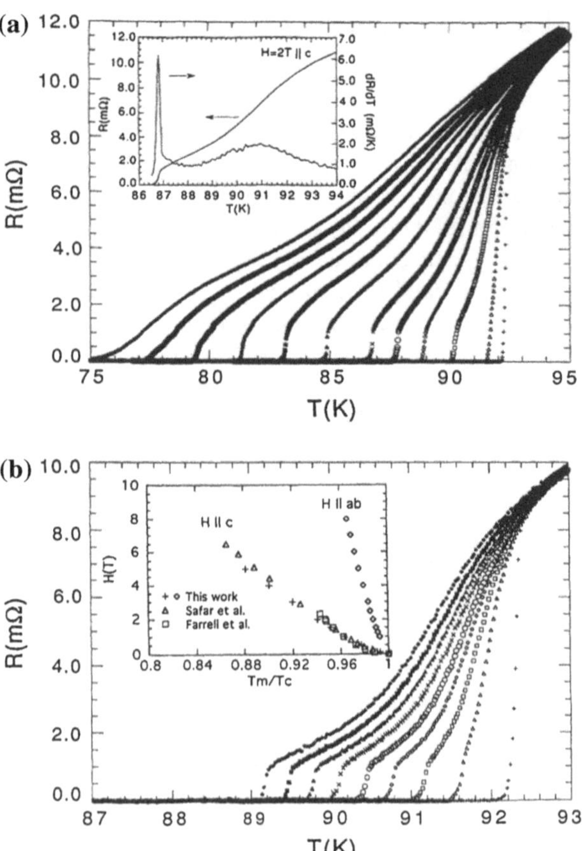

Fig. 5.1 Resistive transition in magnetic fields of (**a**) 0, 0.1, 0.5, 1, 1.5, 2, 3, 4, 5, 6, 7, and 8 T for **H** ∥ *c* and (**b**) 0, 1, 2, 3, 4, 5, 6, 7, and 8 T for **H** ∥ *ab* in a high-quality $YBa_2Cu_3O_7$ single crystal. *Inset* (**a**) Determination of T_m from the peak of dR/dT. *Inset* (**b**) H-T phase diagram of the melting transition [11]

experiments. Strong evidence for the existence of the vortex lattice melting has been drawn from the observation of sharp kinks in the resistive transition of high-quality $YBa_2Cu_3O_7$ (YBCO) single crystals [11–16]. In Fig. 5.1, the resistance show no anomalies at $T_c(H)$ because the flux-flow resistance of the superconducting phase at $T_c(H)$ joins smoothly with the normal state resistance in magnetic fields [17]. Instead, sharp transition, which indicates the existence of the first-order phase transition, occurs at $T_m(H)$ [11]. The resistivity measurements by Safar et al. revealed the reproducible hysteresis, that is the evidence of the first-order phase transition [12]. However, resistivity is not a thermodynamic property, and one may suggest a possibility that the resistive steps indicate only a change in vortex dynamics not a thermodynamic phase transition. In $Bi_2Sr_2CaCu_2O_8$ (BSCCO), where the resistive step is very small [18], a very clear and direct thermodynamic observation of the first-order phase transition has been performed using a micro Hall probe, which could avoid the influence from the intrinsic inhomogeneity of the magnetic induction B, that often masks the nature of first-order phase transition [19]. The vortex lattice melting in BSCCO has been also indicated by disappearance of the neutron diffraction pattern [20] and by the direct observation of the vortex lattice using a scanning Hall probe microscopy [21]. Many other experiments, for instance, magnetization [22, 23] and heat capacity measurements [24, 25], have led well understandings for the nature of first-order melting transition in the high-T_c cuprates.

5.1.4 Vortex Glass

Spacial inhomogeneity also plays an important role for the vortex state in high-T_c cuprates. The defects, which must exist at random in an ordinary compound, lead to the inordinate vortex distribution like a glass rather than a periodic lattice. In this case, the resistivity will disappear at the limit of zero current (non-Ohmic behavior) and the phase transition at $T_g(H)$ between a vortex liquid phase and a vortex glass phase becomes a second-order phase transition, at which the resistivity shows no discontinuous step but gradually decreases in the scaling $\rho(T) \propto (T - T_g)^s$ with decreasing the temperature [26]. Even in very clean YBCO single crystal, the melting transition varies the glass one at high magnetic fields ($\mu_0 H > 5\,\mathrm{T}$) as shown in Fig. 5.1a. Artificially-produced disorder due to the electron irradiation also suppresses the first-order melting transition [27].

5.2 Experiment

5.2.1 Samples

Single crystals of URu_2Si_2 grown by the Czochralski pulling method in a tetra-arc furnace are provided by Dr. Yoshinori Haga group at Japan Atomic Energy Agency. The resistivity shows an exceptionally low residual value of $\rho_0 \simeq 0.5\,\mu\Omega\,\mathrm{cm}$ and large residual resistivity ratio of 670. It attests the highest crystal quality currently achievable.

5.2.2 Transport Measurements

5.2.2.1 Resistivity

The resistivity was measured in a ^{3}He cryostat using a standard four-wire method with an excitation current **J** along the *a* axis. We used an ac resistance bridge LR-700 to measure the resistivity with several excitation currents. The temperature was determined using cernox or ruthenium oxide (RuO_2) resistive thermometers. The

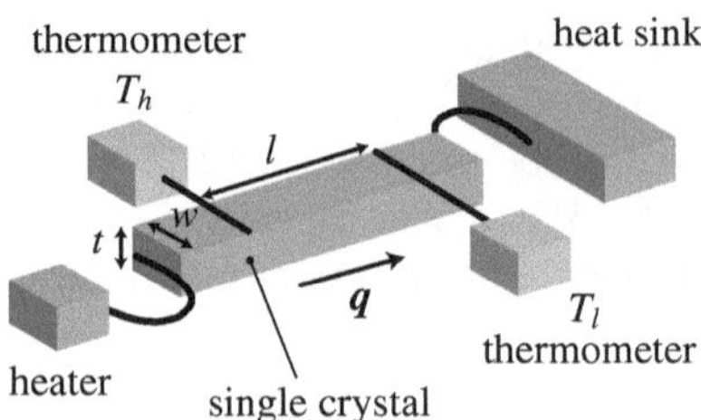

Fig. 5.2 Schematic view of the thermal conductivity probe

temperature was monitored and controlled by CryoCon 62. The magnetic fields were applied for $\mathbf{H} \parallel a$ and $\mathbf{H} \parallel c$ by a 7-T superconducting magnet.

5.2.2.2 Thermal Conductivity

The thermal conductivity was measured in a dilution refrigerator using a standard four-wire steady-state method. In a steady state, the thermal conductivity κ is defined as

$$\mathbf{q} = -\bar{\kappa}\nabla T, \tag{5.11}$$

where $\bar{\kappa}$ is the thermal conductivity tensor, $\mathbf{q}$ is the thermal current density, and ∇T is the temperature gradient. In a one-dimensional system, Eq. (5.11) is expressed as

$$q = -\kappa\frac{dT}{dx}, \tag{5.12}$$

where q is a flux of heat per unit time per unit area and dT/dx is the temperature gradient along the length of the sample. Here we apply $\mathbf{q}$ along the a axis. Figure 5.2 shows a schematic view of the probe for thermal conductivity measurements. We apply the current I to a 1-kΩ resistive heater to make a temperature difference $\Delta T = [T_h(I) - T_l(I)] - [T_h(I = 0) - T_l(I = 0)]$, where T_h and T_l are temperatures at the contacts on the side of the resistive heater and the heat sink, respectively. We then obtain the thermal conductivity as

$$\kappa = \frac{l}{wt}\frac{Q}{\Delta T}, \tag{5.13}$$

where l, w, and t are the distance between gold-wire contacts on a sample, width, and thickness of the sample, respectively. Here $Q = IV$ (V is the measured voltage) is the heat energy per unit time. We used two RuO_2 resistive thermometers to determine T_h and T_l. These resistances were measured by picowatt AVS-47 resistance bridges. We apply I to the heater using Keithley 2400 sourcemeter, and measure the heater voltage V using Keithley 2000 multimeter. Measurements has been performed in a 14-T superconducting magnet.

The thermal conductivity is a powerful probe to know the detailed information of quasiparticles and phonons in superconducting states. Here the thermal conductivity can be expressed as

$$\kappa = \frac{1}{3}Cvl, \tag{5.14}$$

where C, v, and l are the specific heat, mean velocity, and mean free path, respectively. The electron specific heat C_{el} and the phonon specific heat C_{ph} are given by

$$C_{el} = \frac{1}{2}\pi^2 nk_B\left(\frac{T}{T_F}\right), \tag{5.15}$$

$$C_{ph} = \frac{12}{5}\pi^4 Nk_B\left(\frac{T}{\theta_D}\right)^3, \tag{5.16}$$

where n, N, T_F, and θ_D are the electron density, the number of atoms per unit volume, the Fermi temperature, and the Debye temperature, respectively. In clean metals, the thermal current would be dominated by electrons at low temperatures. Now, $l = v_F\tau$, the Fermi energy $\varepsilon_F = k_BT_F = mv_F^2/2$, and the density of states $N(0) = 3n/2\varepsilon_F$, then

$$\kappa \sim \kappa_{el} = \frac{1}{3}C_{el}vl = \frac{\pi^2 nk_B\tau T}{3m} = \frac{\pi^2 k_B^2}{9}N(0)v_F^2\tau T, \tag{5.17}$$

where τ and v_F are scattering time and the Fermi velocity, respectively, and we obtain

$$\frac{\kappa}{T} \propto N(0)v_F^2\tau = N(0)v_F l. \tag{5.18}$$

Using the electrical conductivity $\sigma = ne^2\tau/m$, the ratio between the thermal and electrical conductivities is given by

$$\frac{\kappa}{T\sigma} = \frac{\kappa}{T}\rho = \frac{\pi^2 nk_B^2\tau/3m}{ne^2\tau/m} = \frac{\pi^2}{3}\left(\frac{k_B}{e}\right)^2 = L_0. \tag{5.19}$$

This is the Wiedemann-Franz law and the universal constant $L_0 = 2.45 \times 10^{-8}\,\mathrm{W\Omega/K^2}$ is the Sommerfeld value. When the electronic contribution is enough dominant, the Lorenz number $L \equiv \kappa/T\sigma$ is close to L_0. We can estimate the electronic contribution from the Lorenz number.

5.2.3 Local Magnetization Measurement

The local magnetic induction was measured in a ^{3}He cryostat using a small HgCdTe Hall probe with an active area of $100 \times 100\,\mu\mathrm{m}^2$, which was directly placed on

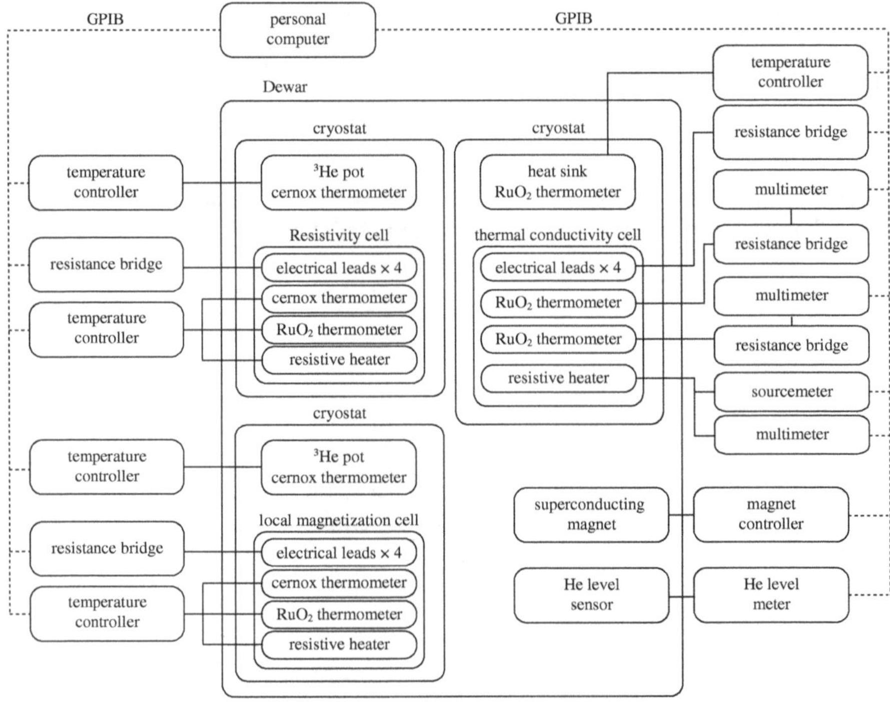

Fig. 5.3 Schematic view of the whole experimental system

the surface of URu_2Si_2. The Hall probe was provided by Dr. Marcin Konczykowski group at Ecole Polytechnique. As also mentioned in previous chapter, the local probe is highly advantageous to sensitively detect the magnetization change in superconductors such as the magnetization jump associated with the first-order vortex-lattice melting phase transitions [19]. An example of such local magnetization measurements will be described below. The local magnetic induction B is determined as $B = R_H^{-1} R$, where R_H is the Hall coefficient of the HgCdTe Hall sensor determined from the blank measurements. The Hall resistance R is measured by an ac resistance bridge Lakeshore 370 and the temperatures are monitored and controlled by CryoCon 62. Figure 5.3 shows the whole experimental system in the present transport study of URu_2Si_2.

5.2.3.1 Local Magnetization of the Heavy-Fermion Compound $CeCoIn_5$

Here we will briefly describe the low-temperature and high-field local magnetization measurement, which is developed for the investigation of unusual superconducting state of the heavy-fermion compound $CeCoIn_5$ ($T_c = 2.3$ K) [28]. This material exhibits a first-order phase transition at H_{c2} at low temperatures [29, 30], in sharp contrast to conventional superconductors, where the transition at H_{c2} is of second order. In $CeCoIn_5$, it has been indicated that Fulde–Ferrell–Larkin–Ovchinnikov (FFLO) state appears in low-temperature and high-field corner of the H-T phase diagram [31, 32]. In the FFLO state, a new pairing ($\mathbf{k}\uparrow, -\mathbf{k}+\mathbf{q}\downarrow$) with finite center-of-mass momenta $\mathbf{q}$ between the Zeeman split parts of the Fermi surface realizes and planar nodes appear periodically perpendicular to the magnetic fields, which segment the flux lines to small vortex pieces [33–35]. It has been widely discussed that the first-order transition and FFLO state appear as a result of the strong Pauli paramagnetism in $CeCoIn_5$. Here, to study detailed information of the superconducting-normal transition in $CeCoIn_5$, we have measured the local magnetic induction by utilizing a micro Hall probe magnetometry with an advantage over the bulk magnetization measurements, by which the nature of the first-order phase transition is often masked by the unavoidable field inhomogeneity inside the superconductors due to the vortex pinning and the surface barrier effects [19].

Figure 5.4 depicts the magnetic field dependence of the Hall resistance $R(H)$ of a miniature Hall probe at 270 mK. Because of the high mobility of the two-dimensional electron gas, the quantum Hall effect is observed. In the plateau region, the Hall sensor is insensitive to the applied field and hence is no use for the measurements. On the other hand, in the region where $R(H)$ shows a steep increase, the sensor is very sensitive to the applied field. To obtain the high sensitivity for the measurements of the magnetic induction, we selected a suitable Hall probe in which $R(H)$ increases steeply at around the upper critical fields of $CeCoIn_5$, as shown in Fig. 5.4.

The insets of Fig. 5.5a, b show the temperature dependence of the Hall resistance $R(T)$ in constant magnetic fields for $\mathbf{H} \parallel c$ and $\mathbf{H} \parallel a$, respectively. At the superconducting transition temperature T_c, $R(T)$ exhibits a step-like behavior at higher fields, while it shows a kink behavior at lower fields, for both field directions. The main panels of Fig. 5.5a, b show the temperature dependence of the local magnetic induction change in the superconducting state relative to the normal state δB for $\mathbf{H} \parallel c$ and $\mathbf{H} \parallel a$, respectively. Here δB is obtained as $\delta B = [dR(H)/dH]^{-1}[R(H) - R_n(H)]$,

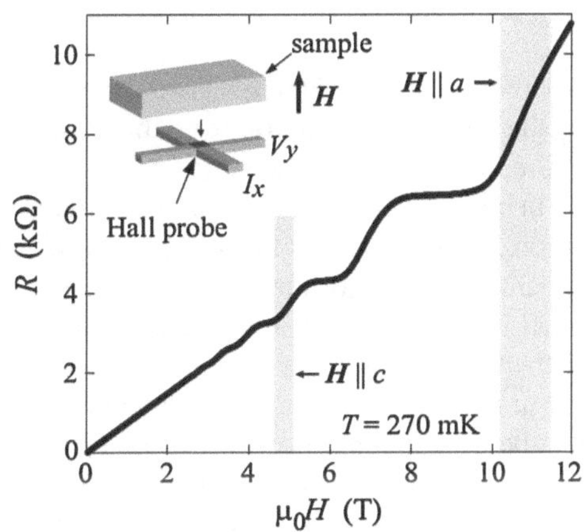

Fig. 5.4 Magnetic field dependence of the Hall resistance of the micro Hall probe at 270 mK. The *hatched* region is used for the measurements for the local magnetic induction of $CeCoIn_5$ for $\mathbf{H} \parallel c$ and $\mathbf{H} \parallel a$. The *inset* illustrates the experimental setup [28]

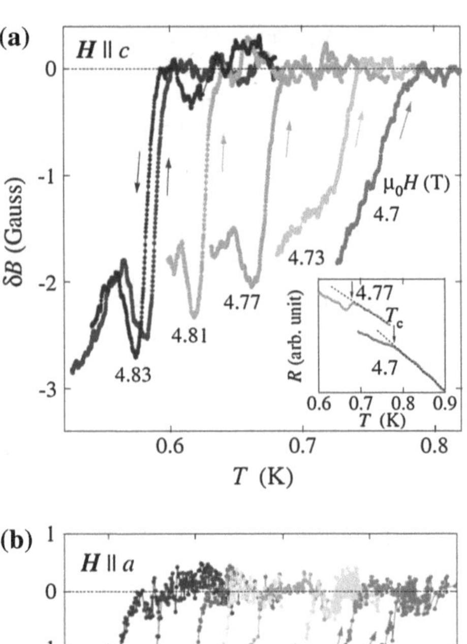

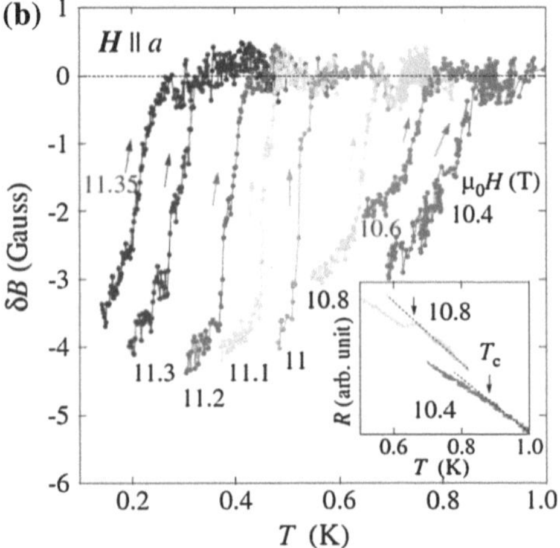

Fig. 5.5 Temperature dependence of the local magnetic induction changes relative to the normal state at various magnetic fields for (**a**) **H** $\parallel c$ and (**b**) **H** $\parallel a$. The *arrows* indicate the field sweep directions. *Insets* Temperature dependence of the Hall resistance. The *arrows* indicate the superconducting transition temperatures. *Dashed line* is a linear fit of $R(T)$ in the normal state [28]

where $R_n(H)$ is the Hall resistance in the normal state. Below T_c, $R_n(H)$ is obtained by the linear extrapolation from the normal state above T_c, as shown by dashed lines in the insets of Fig. 5.5a, b.

In both directions, the local magnetization anomaly is clearly observed at the transition temperatures. At low fields, it shows a kink behavior, indicating a second-order superconducting-to-normal phase transition at H_{c2}. At high magnetic fields, on the other hand, it exhibits a jump associated with a hysteresis behavior at T_c, providing with thermodynamic evidence of first-order phase transition. For **H** $\parallel c$, furthermore, δB jumps to lower values at T_c and decreases after showing an undershoot behavior just below the jump down. It should be noted that this undershoot behavior is only observed at the transition between FFLO and normal state, implying a characteristic feature of such first-order phase transition. We emphasize that this remarkable undershoot behavior of δB has never been observed in the bulk magnetization measurements, though the jump of the magnetization has been reported [30]. It should be also stressed that the observed undershoot behavior is not related to the so-called peak effect, which is often observed in the critical current density in the vicinity of upper critical fields in clean type-II superconductors. First of all, the present experiments have been performed under field cooling conditions, in which the peak effect is hardly observable. Second, the undershoot behavior does not depends on the direction of temperature sweeping (Fig. 5.5a), which is opposite to the case for the peak effect if present. Third, peak effect has never been reported just below upper critical fields even in the field sweep experiments in $CeCoIn_5$.

This unusual local magnetization behavior can be related to the peculiar quasiparticle structure expected in the FFLO state. In the FFLO state, new nodal planes appear in the real space perpendicular to the vortex lines, forming a textured superconducting order parameter [35]. Then new quasiparticle excitations occur in the nodal planes, in addition to those in the vortex lines. According to recent theoretical calculations based on the Eilenberger equation, there is no quasiparticle excitation in the region where the vortex lines intersect with the nodal planes, because of 2π shift of the quantum phase of the order parameter when crossing this region [36]. The paramagnetic moments appear in the nodal planes with the thickness $\sim\xi$ as well as the vortex cores with the radius of $\sim\xi$, where ξ is the coherence length. Then the total magnetization $M_{\rm total}$ is expressed as

$$M_{\rm total} = -M_{\rm s} + M_{\rm plane}\left(\frac{\xi}{\lambda_{\rm FFLO}}\right) + M_{\rm vortex}\left(\frac{\xi}{a_0}\right)^2 - M_{\rm cs}\left(\frac{\xi^3}{a_0^2\lambda_{\rm FFLO}}\right), \quad (5.20)$$

where $\lambda_{\rm FFLO} \sim 2\pi/|\mathbf{q}|$ is the distance between the neighboring nodal planes and $a_0 = \sqrt{\phi_0/H}$ is the inter vortex distance. The first term in the right hand side is the contribution of the superconducting diamagnetism which is negative. Because of the first-order nature of the transition, $-M_{\rm s}$ shows a step-like decrease at $T_{\rm c}$ with lowering temperature. The second and third terms provide positive contributions, which represent the paramagnetic contribution from the quasiparticles excited within the FFLO nodal planes and inside the vortex lines, respectively. The fourth term shows the contribution from the region where the vortex lines intersect with the nodal planes. Because of no quasiparticle excitations in this region, the fourth term provides a negative contribution. When ξ decreases rapidly with decreasing temperature below $T_{\rm c}$, the forth term $\propto \xi^3$, which changes most rapidly, would give rise to an increase of $M_{\rm total}$ in a narrow temperature region just below T_c. This may be an origin of the observed undershoot behavior. In this scenario, the absence of the undershoot behavior for $\mathbf{H} \parallel a$ is explained by the weaker temperature dependence of ξ at higher fields and lower temperatures, compared with that at $\mathbf{H} \parallel c$.

This local-probe measurements have clearly demonstrated an unusual magnetic behavior in $CeCoIn_5$, which has never been observed in other bulk magnetization measurements. In this chapter, we utilize this local technique for detecting magnetization jump associated with a first-order melting transition in URu_2Si_2, in addition to the transport measurements as described above.

5.3 Results

5.3.1 Resistivity and Thermal Conductivity

We first show the temperature dependence of the resistivity measured in magnetic fields for $\mathbf{H} \parallel c$ and $\mathbf{H} \parallel a$ in Figs. 5.6 and 5.7, respectively. In zero field, the resistivity becomes zero at $T_c = 1.4\,\mathrm{K}$ and its transition width is approximately 0.2 K. In contrast, the resistivity in magnetic fields exhibits a kink or very sharp drop at solid arrows, which is even sharper than the transition width in zero field. The temperature derivative of the resistivity $d\rho/dT$ in the insets of Figs. 5.6 and 5.7 demonstrates these anomalies more obviously.

In magnetic fields, $d\rho/dT$ has extremely sharp peak corresponding a sudden drop of the resistivity, while it only shows smooth uprise around $T = 1.5\,\mathrm{K}$ in zero field. Then we determine the kink temperatures by the peak positions of $d\rho/dT$, which are shown by T_m in Figs. 5.6 and 5.7. We will explain the dotted arrows in these figures later. It should be noted that the very large magnetoresistance in the normal state stems from the compensation, i.e. essentially equal number of electrons and holes, $n_e = n_h$ [37].

We next discuss the thermal conductivity κ in the superconducting state of URu_2Si_2, which also provides important information on the vortex state. Figure 5.8a shows the temperature dependence of the thermal conductivity divided by the tem-

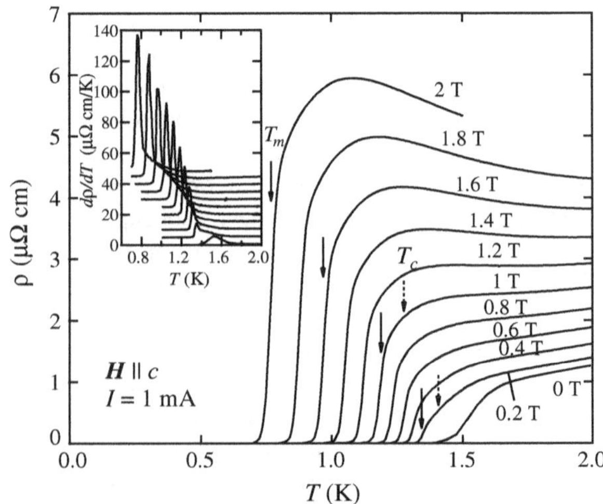

Fig. 5.6 Resistivity ρ as a function of T for $\mathbf{H} \parallel c$. The *solid arrows* indicate the melting transition temperature T_m, which is defined by the peak of $d\rho/dT$. The *dotted arrows* indicate the mean field transition temperatures T_c determined by the cusp of the thermal conductivity as shown in Figs. 5.10 and 5.11. *Inset* The temperature derivative of ρ as a function of T. The $d\rho/dT$ data are shown from 0 to 2 T in increments of 0.2 T with a vertical shift

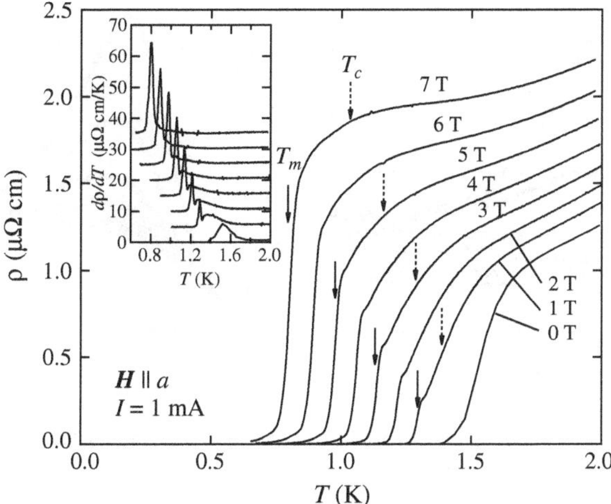

Fig. 5.7 Resistivity ρ as a function of T for $\mathbf{H} \parallel a$. *Inset* shows $d\rho/dT$ as a function of T. The data are shown from 0 to 7 T in increments of 1 T with a vertical shift

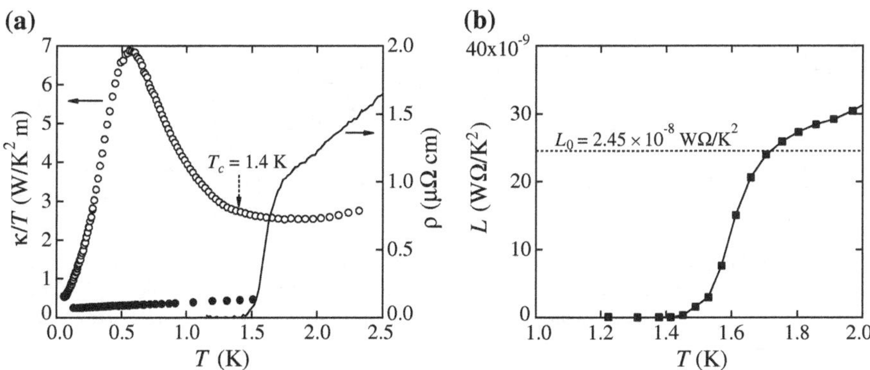

Fig. 5.8 **a** *Left axis* The thermal conductivity divided by the temperature κ/T as a function of T in zero field (*open circles*). *Filled circles* show κ/T data measured in $\mu_0 H = 3\,\text{T}$ for $\mathbf{H} \parallel c$ in the normal state [38]. It indicates the maximum contribution from the phonon. *Right axis* ρ as a function of T in zero field. **b** The Lorenz number $L = \kappa\rho/T$ as a function of T in zero field

perature κ/T. The open circles indicate the κ/T data measured in zero field, which exhibit a notable enhancement below T_c. We now estimate the contributions to the thermal conductivity from the quasiparticles and phonons in the superconducting state. Figure 5.8b depicts the temperature variation of the Lorenz number L. In the normal state, L is very close to the Sommerfeld value L_0. Moreover, the phonon contribution could be estimated by the previous thermal conductivity measurements in magnetic fields above H_{c2}. The filled circles in Fig. 5.8a show κ/T measured at $\mu_0 H = 3\,\text{T} > H_{c2}$ for $\mathbf{H} \parallel c$ [38], which are much smaller than the κ/T in the present study. These indicate that the electronic contribution well dominates over

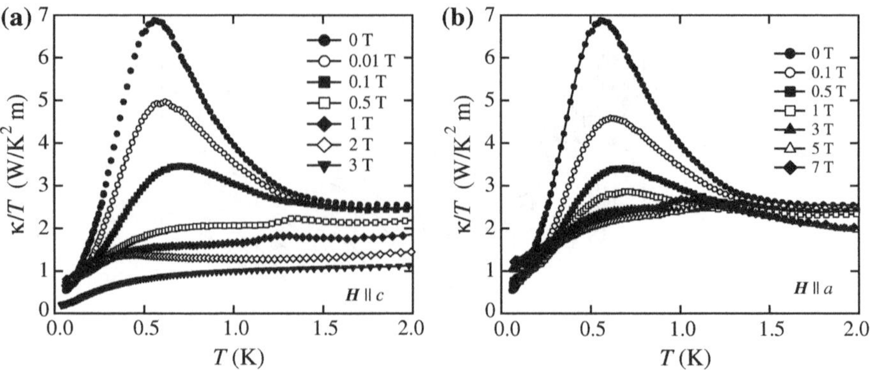

Fig. 5.9 Thermal conductivity divided by the temperature κ/T as a function of T measured in magnetic fields for **a** **H** $\parallel$ c and **b** **H** $\parallel$ a

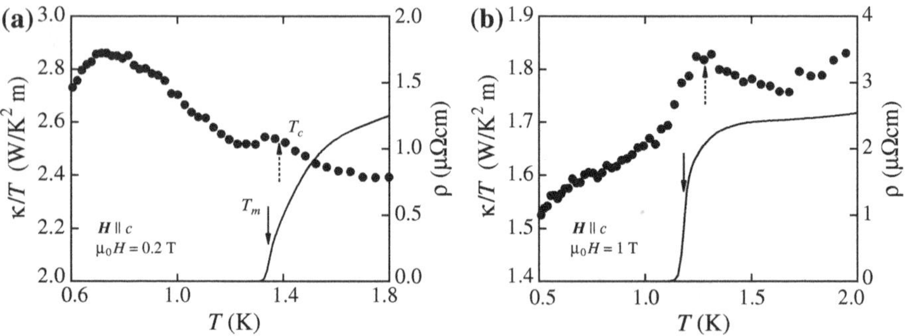

Fig. 5.10 κ/T (*left axis*) and ρ (*right axis*) as a function of T measured in magnetic fields of (**a**) 0.2 T and (**b**) 1 T for **H** $\parallel$ c. The *dotted arrow* indicates the temperature at which κ/T shows a cusp, which correspond to the mean-field transition temperature $T_c(H)$. The *solid arrows* denote the melting temperature T_m

the phonon contribution in this temperature range. As shown in previous section [Eq. (5.18)], the electronic thermal conductivity is described by

$$\frac{\kappa}{T} \sim N(0)v_F l. \tag{5.21}$$

The enhancement of κ/T below T_c is therefore caused by a striking enhancement of l due to the gap formation, which overcomes the reduction of $N(0)$ in the superconducting state, as observed in several strongly correlated systems [39].

The κ/T behavior changes dramatically in magnetic fields as shown in Fig. 5.9. Figure 5.9a, b displays the temperature dependence of κ/T in magnetic fields in a wide temperature range for **H** $\parallel$ c and **H** $\parallel$ a, respectively. In magnetic fields, the enhancement of κ/T in superconducting states is strongly suppressed. Figures 5.10 and 5.11 show the extended views of Fig. 5.9a, b. As the temperature is lowered, κ/T begins to decrease with a distinct cusp shown by the dotted arrows.

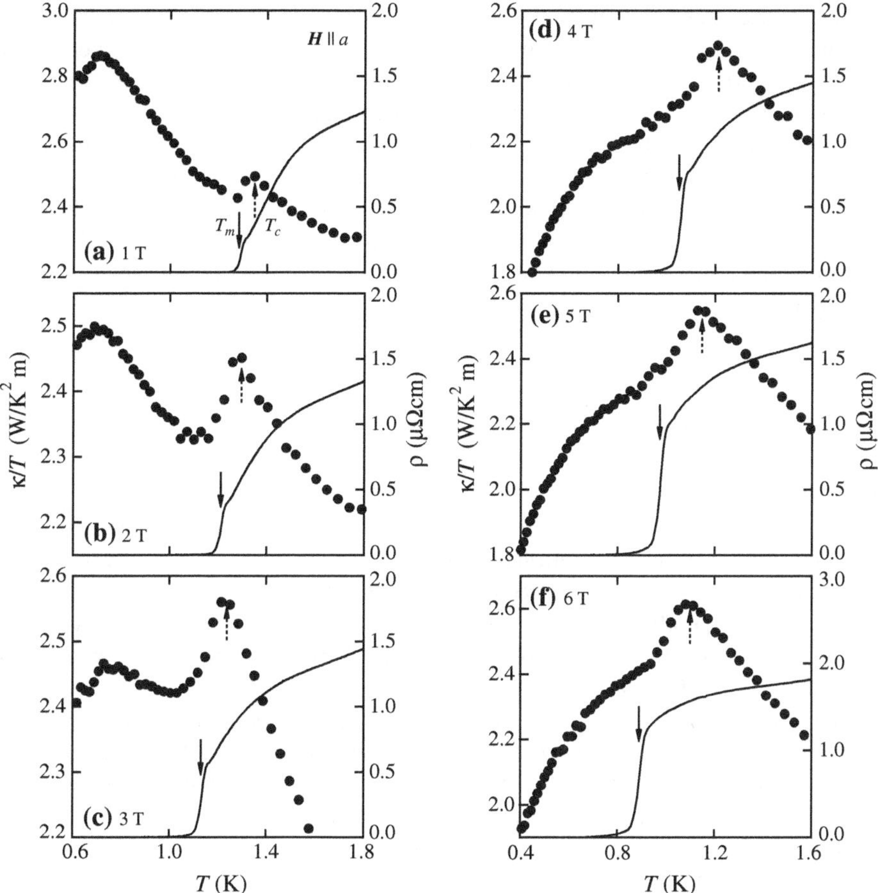

Fig. 5.11 κ/T (*left axis*) and ρ (*right axis*) as a function of T measured in several magnetic fields for $\mathbf{H} \parallel a$. The *dotted arrow* indicates the temperature at which κ/T show a cusp, which correspond to $T_c(H)$. The *solid arrows* denote the melting temperature T_m

According to recent theories [40, 41], thermal conductivity has no fluctuation correction, in contrast to other transport and thermodynamic properties such as resistivity and specific heat which are subject to the fluctuations. Therefore, it is natural to consider that the cusp temperature of κ/T corresponds to the mean-field transition temperature $T_c(H)$. The decrease of κ/T below $T_c(H)$ immediately indicates that l is not enhanced as in zero field and the reduction of $N(0)$ becomes dominantly in Eq. (5.21). Further lowering the temperature brings a second anomaly below which κ/T increase from that extrapolated from high temperatures. This second anomaly is located very close to T_m (solid arrows), indicating that the quasiparticle scattering is notably changed around there, which will be discussed later.

Then we discuss the unusual behavior of the resistivity in magnetic fields as shown in Figs. 5.6 and 5.7, whose transition is even sharper than that in zero field. In Figs. 5.6 and 5.7, the dotted arrows indicate the mean-field transition temperature $T_c(H)$ determined by the cusp of the thermal conductivity. It is obvious that T_c is well above T_m. Here, the features of the resistive transition of URu_2Si_2 bear striking resemblance to that of the vortex lattice melting transition in clean $YBa_2Cu_3O_7$ (YBCO), where the sharp drop of the resistivity is observed in a linear scale at the melting temperature without clear anomaly at $T_c(H)$ (Fig. 5.6). These results could lead to the conclusion that the vortex lattice melting transition takes place at T_m in URu_2Si_2.

5.3.2 E-J Characteristics

We next discuss the electronic field E versus the current density J characteristics in URu_2Si_2. Figure 5.12a, b shows the temperature dependence of the resistivity measured with different excitation currents for $\mathbf{H} \parallel a$ and $\mathbf{H} \parallel c$, respectively. The E-J characteristics is obtained from analyzing the resistivity data. In Fig. 5.13a–d, we show the E-J characteristics measured for several different magnetic fields. Figure 5.14a, b displays the exponent α obtained by the fitting in $E \propto J^{\alpha}$ as a function of T for $\mathbf{H} \parallel a$ and $\mathbf{H} \parallel c$, respectively.

The Ohmic resistivity $\alpha = 1$ is observed at high temperatures and α increases gradually with approaching T_m and shows a steep increase at T_m. This steep increase of α at T_m well agrees with that of high-T_c cuprates [14], also supporting the occurrence of the melting transition at T_m in URu_2Si_2. On the other hand, in term of the part that α is slightly different from unity even above T_c, we need further consideration for the vortex dynamics.

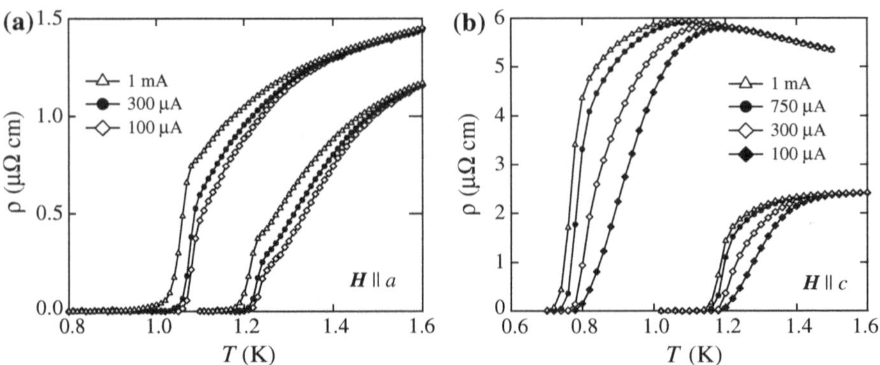

Fig. 5.12 The resistivity as a function of T measured at several excitation currents for (**a**) $\mathbf{H} \parallel a$ and (**b**) $\mathbf{H} \parallel c$

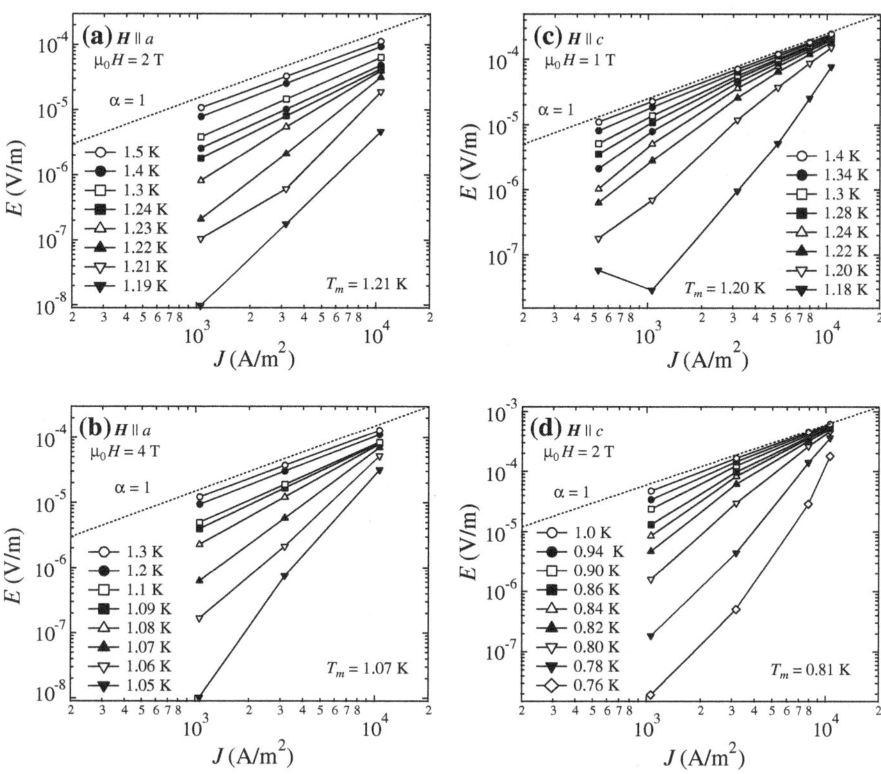

Fig. 5.13 The E-J characteristics around T_m for (**a**, **b**) **H** ∥ a and (**c**, **d**) **H** ∥ c. The *dotted lines* represent the Ohmic behavior $E \propto J$

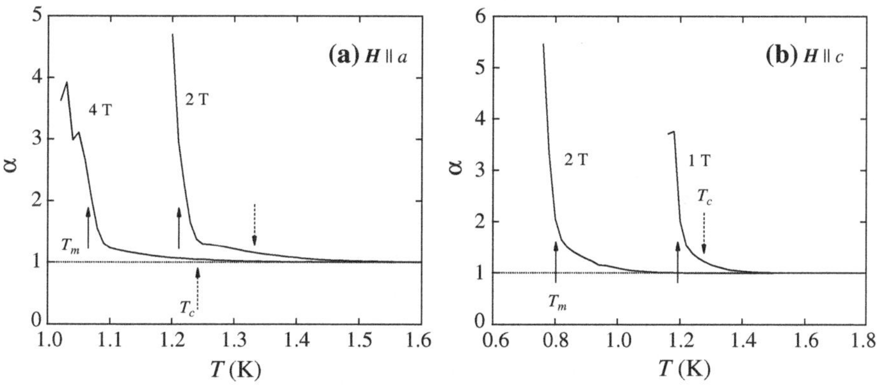

Fig. 5.14 The exponent α as a function of T, which is obtained from the fitting $E \propto J^{\alpha}$

5.3.3 Local Magnetic Induction

Further investigation of the melting transition would be performed by the thermodynamic measurements such as the magnetization and specific heat. In past experiments for the high-T_c cuprates, the local magnetization jump probed by a micro Hall sensor have provided strong evidence of the first-order melting transition [19]. Here, to show a thermodynamic evidence of the melting transition in URu_2Si_2, we have measured the local magnetic induction. Figure 5.15a shows the temperature variations of the Hall resistance R, which is proportional to the local magnetic induction, measured in various constant fields. The change of the local magnetic induction relative to the normal state δB is given by

$$\delta B = R_H^{-1}(R - R_n), \tag{5.22}$$

where R_n is a T-linear Hall resistance extrapolated from the normal state (dashed lines in Fig. 5.15). Figure 5.15b shows the temperature dependence of δB at $\mu_0 H = 0.05\,\mathrm{T}$. We have observed gradual changes of the resistance at around T_m but could not resolve the distinct δB jump at T_m, which could give evidence of the first-order phase transition at T_m.

The entropy and magnetization jumps (ΔS and ΔB) characterizing the first-order phase transition at the vortex lattice melting have been calculated by Dodgson et al. [42]. According to their theory in the small anisotropy model, ΔB is given by

$$\Delta B\ [\mathrm{G}] \approx 1.5 \times 10^{-6} \varepsilon T_m\ [\mathrm{K}] \times (B_m[\mathrm{G}])^{1/2}, \tag{5.23}$$

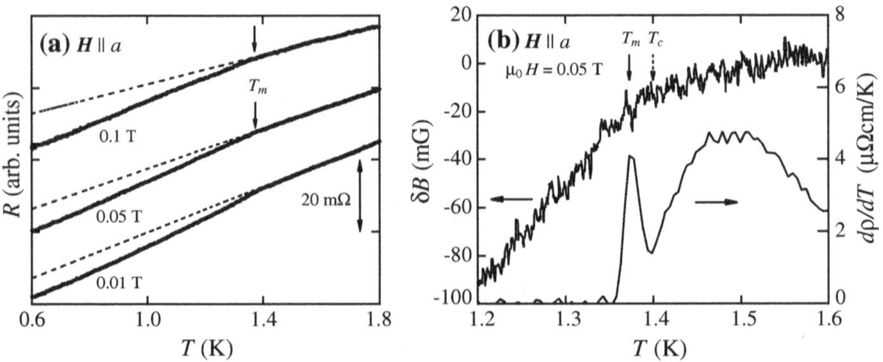

Fig. 5.15 **a** The Hall resistance as a function of T for **H** $\parallel a$. The *dashed lines* are extrapolated from the normal state. The *solid arrows* show the melting temperature T_m determined by the peak of $d\rho/dT$. **b** The local magnetic induction change from the normal state δB and the temperature derivative of the resistivity $d\rho/dT$ as a function of T at $\mu_0 H = 0.05\,\mathrm{T}$. The *solid* and *dotted arrows* show the melting temperature T_m determined by the peak of $d\rho/dT$ and the mean-field transition temperature T_c determined by the cusp of κ/T, respectively

where $\varepsilon \equiv H_{c2}^{a}/H_{c2}^{c}$ is the anisotropy parameter. Using $\varepsilon \approx 4$ for URu_2Si_2, ΔB is estimated to be 1.8×10^{-4} G at $T_m = 1.4$ K and $B_m = 500$ G, which is much smaller than our experimental resolution. From the Clausius-Clapeyron relation, ΔS is also estimated to be ~7.8 μJ/mol K. More precise thermodynamic measurements are highly desirable to obtain evidence of the first-order melting transition in URu_2Si_2.

5.4 Discussion

5.4.1 Melting Transition

We show the H-T phase diagram determined by the present study in Fig. 5.16. The open symbols denote the mean-field H_{c2} lines, which is suggested to be of first-order at very low temperatures (open squares) [37]. The solid square shows the melting transition temperature T_m determined by the clear peak of $d\rho/dT$ as shown in the insets of Figs. 5.6 and 5.7. The most remarkable feature in this phase diagram is

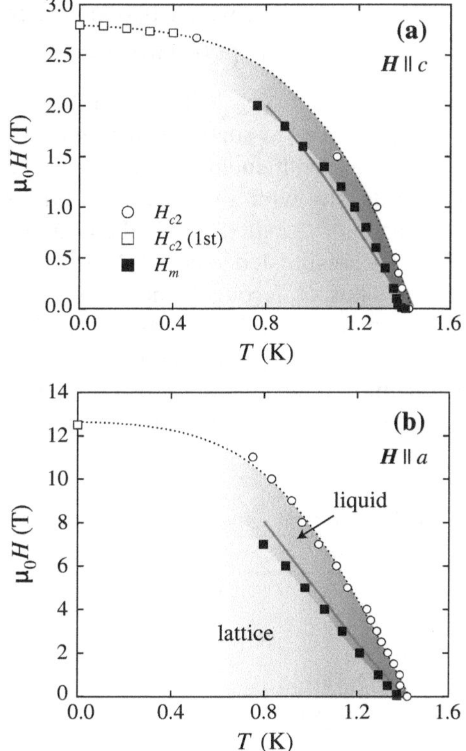

Fig. 5.16 H-T phase diagram of URu_2Si_2 determined by the present study for (**a**) **H** ∥ c and (**b**) **H** ∥ a. *Open symbols* represent the mean field H_{c2} lines. At low temperatures, these lines become first-order (*open squares*) [37]. The *dashed lines* are guides for the eyes. The *solid squares* represent the melting transition H_m, which is fitted by Eq. (5.10) (*thick lines*)

that the melting transition occurs even at very low temperatures and the vortex liquid phase occupies a large portion of the phase diagram compared to conventional low-T_c superconductors. It should be noted that the melting transition line persists to quite high-field regions, at least up to $\sim 0.7 H_{c2}(0)$.

We now discuss the origin of the melting transition in URu_2Si_2. In high-T_c cuprates, extremely large thermal fluctuations extend the critical regions where the mean-field approximation is substantially broken, resulting in the crossover at H_{c2} and the true phase transition line, i.e. the melting transition at T_m. The fundamental parameter that governs the strength of thermal fluctuations is known as the Ginzburg number G_i, which is given by [3]

$$G_i = \frac{1}{2}\left(\frac{\varepsilon k_B T_c}{H_c(0)^2 \xi_a(0)^3}\right)^2 = \frac{1}{2}\left(\frac{8\pi^2 \varepsilon k_B T_c \lambda_a(0)^2}{\phi_0^2 \xi_a(0)}\right)^2, \tag{5.24}$$

where $\lambda_a(0)$ and $\xi_a(0)$ are penetration and coherence lengths in the basal plane at $T = 0\,\mathrm{K}$.

We next quantitatively compare the G_i value of URu_2Si_2 with those of other superconductors. In conventional low-T_c superconductors, G_i ranges from 10^{-11} to 10^{-7}, while in YBCO G_i is as large as $\sim 10^{-2}$ [3]. Table 5.1 summarizes the G_i values in several superconductors. In URu_2Si_2, $\xi_a(0)$ is estimated from the orbital limiting field H_{c2}^{orb}, which is obtained from the initial slope of H_{c2} as $H_{c2}^{orb} = -0.73 \left.\frac{dH_{c2}}{dT}\right|_{T=T_c} T_c$ [45]. Moreover, the μSR measurements report unusually long penetration depth ($\lambda_a = 0.7 - 1\,\mu\mathrm{m}$) [46, 47], giving rise to a large $G_i \sim 3 \times 10^{-4}$ in spite of very low superconducting transition temperature $T_c = 1.4\,\mathrm{K}$. Such a long penetration depth could be naturally understood by the combination of a very low carrier density with a large effective mass in URu_2Si_2. From $n \sim 0.02$ carriers/U atom and $m = 13 m_0$ [37, 48], the London penetration depth $\lambda_L = \sqrt{mc^2/4\pi n e^2}$ in URu_2Si_2 is estimated to be $\sim 1.4\,\mu\mathrm{m}$, which is well consistent with the experimental results. Thus, G_i is roughly increased by five orders of magnitude, leading to a sizable separation of T_m and $T_{c(H)}$, as well as the case of the high-T_c cuprates.

Large G_i leads to the reduction of T_m, extending a vortex liquid region. We fit the melting temperature T_m using Eq. (5.10) with a single fitting parameter $G_i = 3.8 \times 10^{-4}$ as shown by the thick lines in Fig. 5.16. This G_i value is very close

Table 5.1 The Ginzburg number G_i in various superconductors

	T_c [K]	$\xi_a(0)$ [nm]	$\lambda_a(0)$ [nm]	ε	G_i
Nb [43]	9.2	43	35	1	2.2×10^{-11}
$NbSe_2$ [44]	7.1	9.2	140	3	6.1×10^{-7}
MoGe [44]	6.1	5.4	680	1	8.9×10^{-5}
$YBa_2Cu_3O_7$ [3]	93	1.6	140	5	1.0×10^{-2}
$Bi_2Sr_2CaCu_2O_8$ [3]	90	2.5	140	50	4.0×10^{-1}
URu_2Si_2	1.4	7.0	1000	4	3×10^{-4}

to the above estimation. These results lead us to conclude that the exceptionally large thermal fluctuations due to unique electronic state, i.e. very small number of carriers with large effective mass in URu_2Si_2, play an important role even at very low temperatures.

5.4.2 Comparison with Other Superconductors

We next compare to the vortex state in URu_2Si_2 with that in high-T_c cuprates, which have large anisotropy reflecting the 2D crystal structures. In the case of high-T_c cuprates for $\mathbf{H} \parallel c$, the coherence length ξ_c along the c axis is smaller than the distance between the CuO_2 planes, resulting in the 2D pancake vortices weakly connected by the Josephson strings between the layers [49], and the melting and decoupling transitions occur simultaneously [50, 51]. On the other hand, in URu_2Si_2 which has larger coherence length than the lattice constant along the c axis, a 3D melting transition to a line liquid is expected to occur.

Remarkable feature in the phase diagram of URu_2Si_2 is that the melting transition line where the resistivity exhibits sharp drop continues at least up to $\sim 0.7 H_{c2}(0)$. Even in clean YBCO, as shown in Fig. 5.17, $H_m(T)$ ceases to increase at $\sim 0.3 H_{c2}(0)$, above which the glass transition with no sharp resistive drop is observed [27]. We stress that the present ultraclean URu_2Si_2 has remarkably few quenched disorders.

Except in high-T_c cuprates, the melting or glass transition has observed in several superconductors. Figure 5.18 displays the resistive transition and the vortex phase

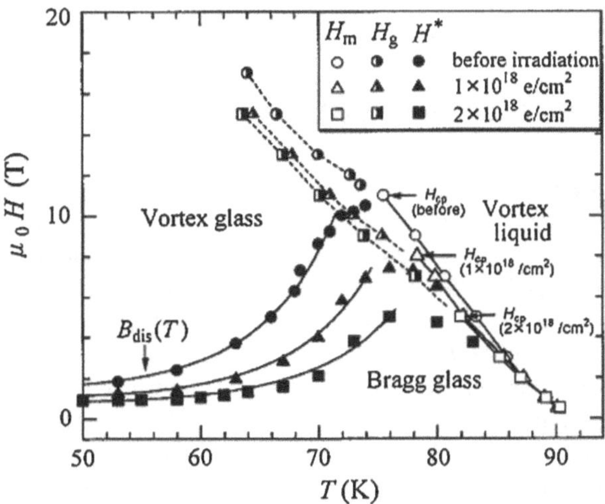

Fig. 5.17 H-T phase diagram of YBCO before and after electron irradiation [27]. The melting transition is suppressed by the electron irradiation. In Bragg glass phase, vortices are weakly pinned but the long-range order is preserved such as the lattice

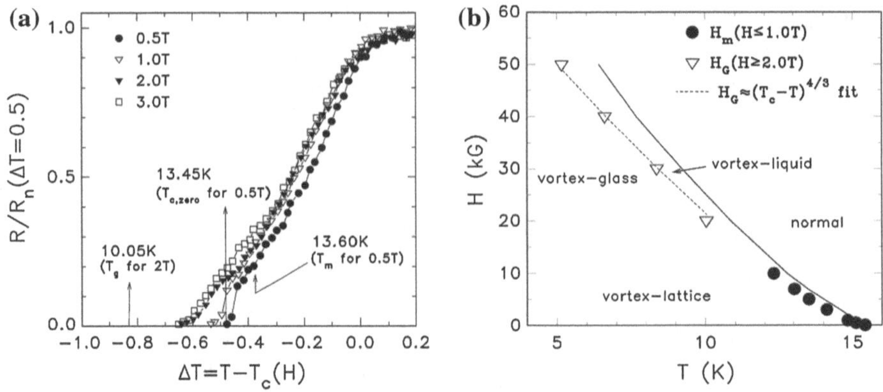

Fig. 5.18 **a** The resistive transition and **b** the H-T phase diagram of YNi_2B_2C [52]

diagram of YNi_2B_2C [52]. We can see that these behaviors are far from that of URu_2Si_2, which has recognizable resistive transition and large portion of vortex liquid phase owing to extreme cleanness and large thermal fluctuation. We also emphasize that the melting transition at sub-Kelvin temperatures has never been observed in other systems.

Recent studies of URu_2Si_2 has revealed quite fascinating phenomena that the first-order phase transition takes place at H_{c2} at very low temperatures, owing to the strong Pauli paramagnetic effect, as shown in Fig. 5.16 (open squares) [37]. The appearance of two distinct first-order-like transition lines with different origins makes this system unprecedented. Exploring the detailed H-T phase diagram at lower temperatures is therefore intriguing.

5.4.3 Quasiparticle Scattering in the Vortex Lattice Phase

We finally discuss the remarkable quasiparticle state in URu_2Si_2. We focus on the unusual κ/T behavior as shown in Figs. 5.6 and 5.7. In magnetic fields, κ/T shows a clear cusp at the mean field transition temperature $T_c(H)$ and then decrease with lowering temperatures. This is because the density of states at the Fermi energy $N(0)$ starts to decrease below T_c. Further lowering temperatures, most interestingly, κ/T begins to increase again at T_m, indicating that the mechanism of the quasiparticle scattering abruptly changed. In our clean URu_2Si_2, however, the quasiparticle mean free path l is estimated to be over 1 μm by comparing the residual resistivity ρ_0 with other samples measured by the dHvA experiments [48]. This is two orders of magnitude longer than the intervortex distance $a_0 \sim 0.04$ μm at $\mu_0 H = 1$ T. In a naive picture, such a long l would not be influenced by the melting transition. Nevertheless κ/T is indeed enhanced below T_m.

Thus, we suggest a possible explanation for this phenomenon, though further studies for understanding this unusual behavior of l are required. We believe that below T_m nearly perfect vortex lattice is formed, in which low-energy quasiparticles are described by the Bloch wave function and are less scattered. The low-energy quasiparticle transport in unconventional superconductors is still a controversial issue. It is suggested that in magnetic fields quasiparticles form Landau levels with a discrete energy spectrum strongly scattered [53]. On the other hand, Franz et al. have shown that low-energy quasiparticles in the vortex lattice are described by the Bloch waves [54], but experimental evidence of their states in the vortex lattice has never been observed until now. Our present studies strongly indicate the realization of quasiparticle Bloch state in the periodic vortex lattice. We also emphasize that the enhancement of κ/T in the vortex solid state has never been observed even in very clean YBCO [55] and $CeCoIn_5$ [56], implying that an ultraclean system is required for the formation of the quasiparticle Bloch state.

It should be noted that recent angle-resolved thermal conductivity measurements have also revealed anomalous κ enhancement in the vortex lattice phase [57]. In the experiment, the thermal conductivity of URu_2Si_2 was measured rotating $\mathbf{H}$ within the basal ab plane with the azimuthal angle ϕ from the heat current direction $\mathbf{q} \parallel a$. They have found that, at $\mu_0 H = 3\,\mathrm{T}$, $\kappa(\phi)$ exhibits a very sharp cusp at $\phi = 0°$, at which the heat current is parallel to the magnetic field direction. Note that such cusp behavior has never been observed in the angle-resolved specific heat measurements [58]. This distinct cusp feature at $\mathbf{H} \parallel \mathbf{q}$ indicates that the quasiparticle mean free path due to vortex scattering l_v scales to the effective intervortex distance along the heat current as $l_v(\phi) = l_v^0/|\sin\phi|$, where l_v^0 is the vortex-scattering mean free path for $\mathbf{H} \perp \mathbf{q}$. This contribution of the thermal conductivity is then described as

$$\kappa_v(\phi) \propto l_{\mathrm{total}}(\phi) = \frac{l_0 l_v(\phi)}{l_0 + l_v(\phi)} = \frac{l_0 l_v^0}{l_v^0 + l_0|\sin\phi|}, \tag{5.25}$$

where l_{total} is the total mean free path defined as $l_{\mathrm{total}}^{-1}(\phi) = l_0^{-1} + l_v^{-1}(\phi)$ and l_0 is the quasiparticle mean free path due to impurities. Now the vortex-scattering mean free path l_v^0 obtained from the fitting of $\kappa(\phi)$ data using $\kappa(\phi) = \kappa_0 + \kappa_{2\phi} + \kappa_v(\phi)$ (κ_0 is a ϕ-independent term and $\kappa_{2\phi} \propto \cos 2\phi$) well exhibits a $B^{-1/2}$ dependence [57], indicating that l_v^0 is well scaled by the intervortex distance a_0. On the other hand, l_v^0 value is one order longer than a_0, indicating a weak quasiparticle scattering in the vortex lattice phase. This is again a contrasting result with that of the heavy fermion $CeCoIn_5$, in which the quasiparticle mean free path becomes comparable to a_0 immediately after entering the mixed state [56].

5.5 Summary

In the present study, we have performed the resistivity, thermal conductivity, and magnetic induction measurements to investigate the nature of vortex state and the quasiparticle structure in URu_2Si_2. We first report the vortex lattice melting transition in URu_2Si_2 from the transport measurements, and summarize the remarkable features as follows.

- We used extremely clean URu_2Si_2 single crystal, which marks large residual resistivity ratio $RRR = 670$ and has huge $\omega_c\tau$ values. Owing to this clean system, we could observe the quite sharp resistive transition even at high magnetic fields, and observe no broad glass transition up to $\sim 0.7H_{c2}$ in contrast to high-T_c cuprate systems.
- We have determined the melting temperature T_m and the mean-field transition temperature T_c line by sharp peak of $d\rho/dT$ and the cusp of κ/T due to the decrease of $N(0)$ in superconducting state, respectively. We find that the melting transition at T_m occurs well below T_c.
- E-J characteristics show steep increase at T_m, also supporting the melting transition, although we could not resolve the magnetic induction jump, which is estimated to be much smaller than our experimental resolution.
- The origin of the melting transition is large thermal fluctuations due to the unusual electronic properties of URu_2Si_2, namely the small number of carriers with heavy mass.
- The unusual enhancement of κ/T below T_m indicates the formation of coherent quasiparticle Bloch state in the vortex lattice phase, which has never observed in other clean systems.

The present results show the unique vortex and quasiparticle states in URu_2Si_2, and demonstrate that heavy fermion superconductors may provide a new playground to study novel vortex matter physics as well as quasiparticle dynamics in the vortex state of type-II superconductors.

References

1. C.J. Lobb, Phys. Rev. B **36**, 3930 (1987)
2. A. Kapitulnik, M.R. Beasley, C. Castellani, C. Di Castro, Phys. Rev. B **37**, 537 (1988)
3. G. Blatter, M.V. Feigel'man, V.B. Geshkenbein, A.I. Larkin, V.M. Vinokur, Rev. Mod. Phys. **66**, 1125 (1994)
4. A. Houghton, R.A. Pelcovits, A. Sudbø, Phys. Rev. B **40**, 6763 (1989)
5. D.R. Nelson, Phys. Rev. Lett. **60**, 1973 (1988)
6. D.R. Nelson, H.S. Seung, Phys. Rev. B **39**, 9153 (1989)
7. E.H. Brandt, Phys. Rev. Lett. **63**, 1106 (1989)
8. M. Tinkham, *Introduction to Superconductivity* (McGraw-Hill, London, 1996)
9. E. Brézin, D.R. Nelson, A. Thiaville, Phys. Rev. B **31**, 7124 (1985)
10. A. Larkin, A. Varlamov, *Theory of Fluctuations in Superconductors* (Oxford Univrersity Press, Oxford, 2005)
11. W.K. Kwok, S. Fleshler, U. Welp, V.M. Vinokur, J. Downey, G.W. Crabtree, Phys. Rev. Lett. **69**, 3370 (1992)
12. H. Safar, P.L. Gammel, D.A. Huse, D.J. Bishop, J.P. Rice, D.M. Ginsberg, Phys. Rev. Lett. **69**, 824 (1992)
13. H. Safar, P.L. Gammel, D.A. Huse, D.J. Bishop, W.C. Lee, J. Giapintzakis, D.M. Ginsberg, Phys. Rev. Lett. **70**, 800 (1993)

14. W.K. Kwok, J. Fendrich, S. Fleshler, U. Welp, J. Downey, G.W. Crabtree, Phys. Rev. Lett. **72**, 1092 (1994)
15. W. Jiang, N.-C. Yeh, D.S. Reed, U. Kriplani, F. Holtzberg, Phys. Rev. Lett. **74**, 1438 (1994)
16. K. Shibata, T. Nishizaki, T. Sasaki, N. Kobayashi, Phys. Rev. B **66**, 214518 (2002)
17. R. Ikeda, T. Ohmi, T. Tsuneto, J. Phys. Soc. Jpn. **58**, 1377 (1989)
18. D.T. Fuchs, E. Zeldov, D. Majer, R.A. Doyle, T. Tamegai, S. Ooi, M. Konczykowski, Phys. Rev. B **54**, R796 (1996)
19. E. Zeldov, D. Majer, M. Konczycowski, V.B. Geshkenbein, V.M. Vinokur, H. Shtrikman, Nature **375**, 373 (1995)
20. R. Cubitt, E.M. Forgan, G. Yang, S.L. Lee, D. Mck Paul, H.A. Mook, M. Yethiraj, P.H. Kes, T.W. Li, A.A. Menovsky, Z. Tarnawski, K. Mortensen, Nature **365**, 407 (1993)
21. A. Oral, J.C. Barnard, S.J. Bending, I.I. Kaya, S. Ooi, T. Tamegai, M. Henini, Phys. Rev. Lett. **80**, 3610 (1998)
22. R. Liang, D.A. Bonn, W.N. Hardy, Phys. Rev. Lett. **76**, 835 (1996)
23. U. Welp, J.A. Fendrich, W.K. Kwok, G.W. Crabtree, B.W. Veal, Phys. Rev. Lett. **76**, 4809 (1996)
24. A. Schilling, R.A. Fisher, N.E. Phillips, U. Welp, D. Dasgupta, W.K. Kwok, G.W. Crabtree, Nature **382**, 791 (1996)
25. A. Junod, M. Roulin, J.-Y. Genoud, B. Revaz, A. Erb, E. Walker, Physica C **275**, 245 (1997)
26. R.H. Koch, V. Foglietti, W.J. Gallagher, G. Koren, A. Gupta, M.P. Fisher, Phys. Rev. Lett. **63**, 1511 (1989)
27. T. Nishizaki, T. Naito, S. Okayasu, A. Iwase, N. Kobayashi, Phys. Rev. B **61**, 3649 (2000)
28. R. Okazaki, H. Shishido, T. Shibauchi, M. Konczykowski, A. Buzdin, Y. Matsuda, Phys. Rev. B **76**, 224529 (2007)
29. K. Izawa, H. Yamaguchi, Y. Matsuda, H. Shishido, R. Settai, Y. Onuki, Phys. Rev. Lett. **87**, 057002 (2001)
30. T. Tayama, A. Harita, T. Sakakibara, Y. Haga, H. Shishido, R. Settai, Y. Onuki, Phys. Rev. B **65**, 180504(R) (2002)
31. H.A. Radovan, N.A. Fortune, T.P. Murphy, S.T. Hannahs, E.C. Palm, S.W. Tozer, D. Hall, Nature **25**, 51 (2003)
32. A. Bianchi, R. Movshovich, C. Capan, P.G. Pagliuso, J.L. Sarrao, Phys. Rev. Lett. **91**, 187004 (2003)
33. P. Fulde, R.A. Ferrell, Phys. Rev. **135**, A550 (1964)
34. A. I. Larkin, Y. N. Ovchinnikov, Zh. Eksp. Teor. Fiz. **47**, 1136 (1964) (Sov. Phys. JETP 20, 762 (1965))
35. Y. Matsuda, H. Shimahara, J. Phys. Soc. Jpn. **76**, 051005 (2007)
36. T. Mizushima, K. Machida, M. Ichioka, Phys. Rev. Lett. **95**, 117003 (2005)
37. Y. Kasahara, T. Iwasawa, H. Shishido, T. Shibauchi, K. Behnia, Y. Haga, T.D. Matsuda, Y. Onuki, M. Sigrist, Y. Matsuda, Phys. Rev. Lett. **99**, 116402 (2007)
38. K. Behnia, D. Jaccard, J. Sierro, P. Lejay, J. Flouquet, Physica C **196**, 57 (1992)
39. Y. Kasahara, Y. Shimono, T. Shibauchi, Y. Matsuda, S. Yonezawa, Y. Muraoka, Z. Hiroi, Phys. Rev. Lett. **96**, 247004 (2006)
40. D.R. Niven, R.A. Smith, Phys. Rev. B **66**, 214505 (2002)
41. S. Vishveshwara, M.P.A. Fisher, Phys. Rev. B **64**, 134507 (2001)
42. M.J.W. Dodgson, V.B. Geshkenbein, H. Nordborg, G. Blatter, Phys. Rev. Lett. **80**, 837 (1998)
43. D.K. Finnemore, T.F. Stromberg, C.A. Swenson, Phys. Rev. **149**, 231 (1966)
44. N. Kokubo, T. Asada, K. Kadowaki, K. Takita, T.G. Sorop, P.H. Kes, Phys. Rev. B **75**, 184512 (2007)
45. D. Saint-James, G. Sarma, E.J. Thomas, *Type II Superconductivity* (Pergamon Press, Oxford, 1969)
46. E.A. Knetsch, A.A. Menovsky, G.J. Nieuwenhuys, J.A. Mydosh, A. Amato, R. Feyerherm, F.N. Gygaxb, A. Schenck, R.H. Heffner, D.E. MacLaughlin, Physica B **186**, 300 (1993)
47. A. Amato, Rev. Mod. Phys. **69**, 1119 (1997)

48. H. Ohkuni, Y. Inada, Y. Tokiwa, K. Sakurai, R. Settai, T. Honma, Y. Haga, E. Yamamoto, Y. Onuki, H. Yamagami, S. Takahashi, T. Yanagisawa, Philo. Mag. B **79**, 1045 (1999)
49. J.R. Clem, Phys. Rev. B **43**, 7837 (1991)
50. T. Shibauchi, T. Nakano, M. Sato, T. Kisu, N. Kameda, N. Okuda, S. Ooi, T. Tamegai, Layer phase coherence in vortex matter phases of $Bi_2Sr_2CaCu_2O_{8+y}$. Phys. Rev. Lett. **83**, 1010 (1999)
51. M.B. Gaifullin, Y. Matsuda, N. Chikumoto, J. Shimoyama, K. Kishio, Phys. Rev. Lett. **84**, 2945 (2000)
52. Mi-Ock Mun, Sung-Ik Lee, W.C. Lee, P.C. Canfield, B.K. Cho, D.C. Johnston, Phys. Rev. Lett. **76**, 2790 (1996)
53. A. S. Melnukov, J. Phys. Condens. Matt. **21**, 4219 (1999)
54. M. Franz, Z. Tešanović, Phys. Rev. Lett. **84**, 554 (2000)
55. R. Ocaña, A. Taldekov, P. Esquinazi, Y. Kopelevich, J. Low Temp, Phys. **123**, 181 (2001)
56. Y. Kasahara, Y. Nakajima, K. Izawa, Y. Matsuda, K. Behnia, H. Shishido, R. Settai, Y. Onuki, Phys. Rev. B **72**, 214515 (2005)
57. Y. Kasahara, H. Shishido, T. Shibauchi, Y. Haga, T.D. Matsuda, Y. Onuki, Y. Matsuda, New J. Phys. **11**, 055061 (2009)
58. K. Yano, T. Sakakibara, T. Tayama, M. Yokoyama, H. Amitsuka, Y. Homma, P. Miranovic, M. Ichioka, Y. Tsutsumi, K. Machida, Phys. Rev. Lett. **100**, 017004 (2008)

Chapter 6
Conclusions

The heavy-fermion compound URu_2Si_2 has attracted much interest because of the so-called "hidden-order" transition at $T_0 = 17.5$ K, whose order parameter is still a mystery, as well as unconventional superconducting state below $T_c = 1.4$ K embedded in the hidden-order state [1–3]. In present study, we have investigated the thermodynamic and transport properties in the hidden-order and the superconducting phases.

On the nature of hidden-order phase, we investigate the out-of-plane and in-plane magnetic torque with developing sensitive torque measurement system [4]. The in-plane torque measured with precisely rotating field in the tetragonal *ab* plane shows four-fold oscillation behavior and no two-fold oscillation in the paramagnetic phase above $T_0 = 17.5$ K, which is entirely expected in the tetragonal crystalline system. Most remarkably, on the other hand, the in-plane torque exhibits notable two-fold oscillation which follows $\cos 2\phi$ form below T_0, indicating that an orthorhombicity with $\chi_{ab} \neq 0$ appears in the hidden-order phase. Our results then capture a new essential feature of the hidden order parameter in URu_2Si_2, i.e. the low-temperature ordered phase is an electronic "nematic" state that breaks four-fold rotational symmetry of the tetragonal crystal structure. Our results also impose strong constraints on theoretical models which attempt to explain the hidden order. Recently, several theoretical models such as degenerate E-type quadrupolar order [5] or E^--type dotriakontapolar order [6], which are based on our observation of broken four-fold symmetry, have been proposed.

Such four-fold rotational symmetry breaking in the hidden-order phase should affect the superconducting state of this material, which emerges only in the hidden-order phase [7]. We therefore develop a precise measurement of the lower critical field, which is closely related to the superfluid density, by utilizing the miniature Hall-sensor array [8]. The temperature dependence of the lower critical field in URu_2Si_2 is found to be highly anomalous, highlighted by non-BCS-type temperature dependence as well as the kink structure [9]. This phenomenon is possibly attributed to the

R. Okazaki, *Hidden Order and Exotic Superconductivity in the Heavy-Fermion Compound URu_2Si_2*, Springer Theses, DOI: 10.1007/978-4-431-54592-7_6,

multicomponent superconducting order parameter in URu_2Si_2, which may possess two different transition temperatures due to the broken four-fold rotational symmetry.

We have also investigated the thermal- and electro-transport properties of the vortex mixed state in URu_2Si_2 and found an intriguing phase transition phenomenon inside the superconducting phase [10]. The phase transition we newly found in this material is a melting transition between vortex lattice and liquid, reminiscent of the melting transition observed in high-T_c cuprates with large thermal fluctuations. We find that the thermal fluctuation is highly enhanced by the unusual electronic state of URu_2Si_2, namely large effective mass and very low carrier density. Also, we have observed the anomalous enhancement of the quasiparticle mean free path below the melting temperature, indicating realization of a coherent quasiparticle Bloch state in the vortex lattice phase.

References

1. T.T.M. Palstra, A.A. Menovsky, J. van den Berg, A.J. Dirkmaat, P.H. Kes, G.J. Nieuwenhuys, J.A. Mydosh, Phys. Rev. Lett. **55**, 2727 (1985)
2. M.B. Maple, J.W. Chen, Y. Dalichaouch, T. Kohara, C. Rossel, M.S. Torikachvili, M.W. McElfresh, J.D. Thompson, Phys. Rev. Lett. **56**, 185 (1986)
3. W. Schlabitz, J. Baumann, B. Pollit, U. Rauchschwalbe, H.M. Mayer, U. Ahlheim, C.D. Bredl, Z. Phys. B **62**, 171 (1986)
4. R. Okazaki, T. Shibauchi, H.J. Shi, Y. Haga, T.D. Matsuda, E. Yamamoto, Y. Onuki, H. Ikeda, Y. Matsuda, Science **331**, 439 (2011)
5. P. Thalmeier, T. Takimoto, Phys. Rev. B **83**, 165110 (2011)
6. H. Ikeda, M.-T. Suzuki, R. Arita, T. Takimoto, T. Shibauchi, Y. Matsuda, Nat. Phys. **8**, 528 (2012)
7. H. Amitsuka, K. Matsuda, I. Kawasaki, K. Tenya, M. Yokoyama, C. Sekine, N. Tateiwa, T.C. Kobayashi, S. Kawarazaki, H. Yoshizawa, J. Magn. Magn. Mater. **310**, 214 (2007)
8. R. Okazaki, M. Konczykowski, C.J. van der Beek, T. Kato, K. Hashimoto, M. Shimozawa, H. Shishido, M. Yamashita, M. Ishikado, H. Kito, A. Iyo, H. Eisaki, S. Shamoto, T. Shibauchi, Y. Matsuda, Phys. Rev. B **79**, 064520 (2009)
9. R. Okazaki, M. Shimozawa, H. Shishido, M. Konczykowski, Y. Haga, T.D. Matsuda, E. Yamamoto, Y. Onuki, Y. Yanase, T. Shibauchi, Y. Matsuda, J. Phys. Soc. Jpn. **79**, 084705 (2010)
10. R. Okazaki, Y. Kasahara, H. Shishido, M. Konczykowski, K. Behnia, Y. Haga, T.D. Matsuda, Y. Onuki, T. Shibauchi, Y. Matsuda, Phys. Rev. Lett. **100**, 037004 (2008)

Curriculum Vitae

Ryuji Okazaki

Current Affiliation
Department of Physics, Nagoya University
Chikusa, Nagoya 464-8602, Japan
Tel. & Fax: +81-52-789-5105
e-mail: okazaki.ryuji@cc.nagoya-u.ac.jp
URL: http://vlab-nu.jp/en/e_home.html

Education

- April 2013: Ph.D. degree, Kyoto University
  Thesis advisor: Professor Yuji Matsuda
  Thesis topic: Hidden order and exotic superconductivity in the heavy-fermion compound URu_2Si_2
- April 2008—September 2010: Ph.D., Department of Physics, Graduate School of Science, Kyoto University
- April 2006—March 2008: M.Sc., Department of Physics, Graduate School of Science, Kyoto University
- April 2002—March 2006: B.Sc., Faculty of Science, Kyoto University

Experience

- October 2010—present: Assistant Professor in Department of Physics, Nagoya University
- April 2008—September 2010: Research Fellow of the Japan Society for the Promotion of Science

R. Okazaki, *Hidden Order and Exotic Superconductivity in the Heavy-Fermion Compound URu_2Si_2*, Springer Theses, DOI: 10.1007/978-4-431-54592-7,

Research Interests

My research interest is various kinds of phase transition phenomena in correlated electron systems such as transition-metal oxides or heavy-fermion compounds. In these materials, diverse ordered states are realized owing to multiple degrees of freedom of electrons. I am also interested in a wide variety of functional properties in such systems, such as thermoelectric and photovoltaic effects, multicentric behaviors and superconductivity, which is being explored in a border of basic and applied physics. Recently I am interested in transport phenomena in highly non-equilibrium condition, which include nonlinear conduction phenomena in high electric field and a thermoelectric transport under illumination.

Zeitfracht Medien GmbH
Ferdinand-Jühlke-Straße 7
99095 Erfurt, Deutschland
produktsicherheit@kolibri360.de